Graph Theory

Euler's Rich Legacy

CONTEMPORARY APPLIED MATHEMATICS

Graph Theory

Euler's Rich Legacy

Wayne Copes
Clifford Sloyer
Robert Stark
William Sacco

JANSON PUBLICATIONS, INC. Providence, Rhode Island

Material based on work supported by the National Science Foundation and produced by the Committee on Enrichment Modules, Department of Mathematical
Sciences, University of Delaware, and Tri-Analytics, Bel Air, Maryland.

Contents

The authors wish to thank Jane Melville and John Jameson, who helped to produce this monograph.

Introduction

Graph theory is an important topic in *modern* mathematics, yet its roots are in the eighteenth century. Its basic concepts are simple and powerful, providing models and solutions for many real world problems. On the other hand, several simply stated problems in graph theory have remained unsolved for generations.

In this monograph we introduce you to the history and application of graph theory. We begin by asking: What is a graph?

In elementary algebra, the word graph refers to a set of points whose (x, y) coordinates satisfy an equation such as $y = 2x + 3$ or $y = x^2 + 4$.

But the graphs you will study here are different. In this monograph, a graph is defined as: *A set of points, or vertices, and a set of lines, called edges, which connect pairs of vertices.** Vertices are sometimes called *nodes*.

Several graphs are shown in Figure 1. Note that edges need not be straight lines.

Graph A has 4 vertices and 3 edges. The edges connect vertices 1 and 2, vertices 1 and 4, and vertices 3 and 4. Graph B contains 3 vertices and *no* edges. Graphs without edges are called nonadjacent graphs.

1. List the number of vertices and edges for graphs C, D, F, and H.

	Number of Vertices	**Number of Edges**
Graph C	_____	_____
Graph D	_____	_____
Graph F	_____	_____
Graph H	_____	_____

So far we have defined graph, edge, and vertex. We need to add a few more terms to our vocabulary.

If two vertices are joined by an edge, they are said to be *adjacent*. In Graph A of Figure 1, vertices 1 and 2 are adjacent. So are vertices 1 and 4 and vertices 3 and 4. Vertices 1 and 3 are not adjacent. Neither are vertices 2 and 3 nor vertices 2 and 4.

2. (a) In Graph C, which pairs of vertices are adjacent? _____

   (b) Which vertices are not adjacent? _____

A graph is said to be *complete* if every pair of vertices is adjacent. That is, a complete graph contains all possible edges. A graph with two vertices can have at most one edge. Therefore, this graph is complete.

●━━━━━━━●

---

*The definitions of all graph related terms used in this monograph are in Appendix I.

A graph with three vertices can have at most three edges.

Such a graph is complete.

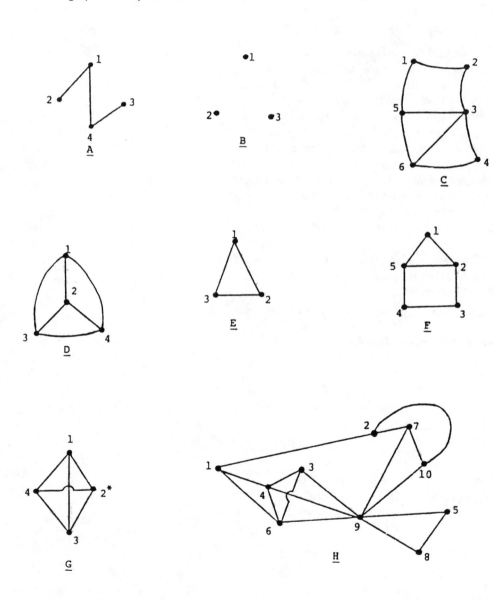

Figure 1

* Note the half-circle on the edge connecting vertices 2 and 4.
It signifies that there is no vertex at the crossing of edges 2–4
and 1–3. This convention is also frequently used in electrical
circuit diagrams.

3. Which graphs of Figure 1 are complete? _____

4. How many edges are in complete graphs with

    (a) 4 vertices? _____

    (b) 5 vertices? _____

    (c) N vertices? _____

Complete graphs are usually denoted by K_N, where the subscript N represents the number of vertices. Thus, K_3 and K_5 represent complete graphs with 3 and 5 vertices, respectively.

A graph is said to be *bipartite* if its vertices can be separated into *two* distinct sets so that each edge has one endpoint in each set.

Consider the complete graph, K_2.

K_2 is bipartite because its vertices can be separated into two distinct sets {1} and {2} so that each edge has one endpoint in each set.

But consider the graph below.

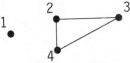

This is not a bipartite graph because, for example, if vertex 2 is in one set, then vertices 3 and 4 must be in the other because they are adjacent to 2. But 3 and 4 are adjacent so they cannot be in the same set.

5. Are the following graphs bipartite? For each bipartite graph list the two sets of vertices which define it.

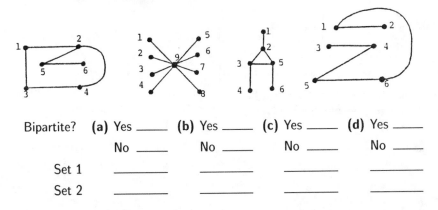

Bipartite?	**(a)** Yes _____	**(b)** Yes _____	**(c)** Yes _____	**(d)** Yes _____
	No _____	No _____	No _____	No _____
Set 1	_____	_____	_____	_____
Set 2	_____	_____	_____	_____

A bipartite graph is said to be *complete* if it contains all possible edges between pairs of vertices in which one vertex comes from each of the two sets. These graphs are denoted by $K_{M,N}$ where M and N are the numbers of vertices in the two sets of vertices. For example, $K_{2,2}$ is shown below.

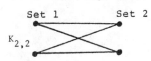

Note that no edge connects two vertices both in set 1 or both in set 2 and that every possible edge connecting a vertex in set 1 with a vertex in set 2 is present.

To emphasize the distinction between a complete graph and a complete bipartite graph, K_4 and $K_{2,2}$ are shown below. Both have 4 vertices. K_4 has 6 edges and $K_{2,2}$ has only 4 edges.

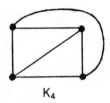

K_4

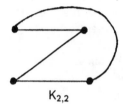

$K_{2,2}$

6. How many edges are in

   (a) $K_{3,3}$? _____

   (b) $K_{5,3}$? _____

   (c) $K_{M,N}$? _____

7. Draw $K_{5,3}$ below. The two sets of vertices are provided.

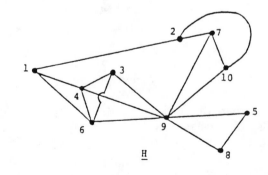

The *degree* of a vertex is the number of edges emerging from it. Graph C has three vertices of degree 2 (vertices 1, 2, and 4), two vertices of degree 3 (vertex 5 and vertex 6), and one vertex of degree 4 (vertex 3).

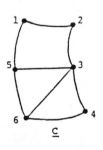

C

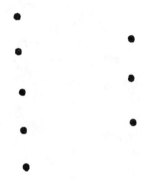

H

8. List the degrees of the vertices in Graph H at the bottom of page 4.

Vertex	1	2	3	4	5	6	7	8	9	10
Degree	___	___	___	___	___	___	___	___	___	___

Now you know what graphs are and a few of their basic properties. But what started the study of graphs?

Systematic study of graph theory began in the 18th century with the famous problem of the seven bridges of Königsberg, a university town in eastern Europe. These bridges connected the banks of the Pregel river and two islands, as shown below.

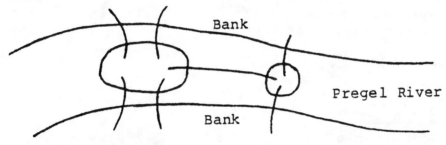

Students and teachers on the mathematics faculty used to wrestle with the question whether it was possible to take a stroll, starting and ending at the same point, which crosses each bridge exactly once. The famous mathematician Leonhard Euler* showed it could not be done, and his proof began the formal development of graph theory. More detailed accounts of this problem are given in the cited references.

The following sections present several topics in graph theory and their applications.

---

*See Appendix II for a short biographical sketch. Euler is pronounced "oiler".

Part I
CHROMATIC NUMBER

Table 1 is a matrix which indicates the subjects taken by various students. If student i is taking subject j then a "1" appears in the matrix in the (i,j)th position. Otherwise a "0" appears. For example, student 5 is taking subject 2 since there is a "1" in position (5,2). Student 5 is also taking subjects 7 and 10.

Table 1. Student-Subject Matrix

Subject

	1	2	3	4	5	6	7	8	9	10
1	1	0	0	1	0	1	0	0	0	0
2	1	1	1	0	0	0	0	0	0	0
3	0	0	1	1	0	1	0	0	0	0
4	0	0	0	0	1	0	0	1	1	0
5	0	1	0	0	0	0	1	0	0	1
6	1	0	0	1	0	1	0	0	0	0
7	1	1	1	0	0	0	0	0	0	0
8	0	0	1	1	0	1	0	0	1	0
9	0	0	0	0	1	0	0	1	0	0
10	0	1	0	0	0	0	1	0	1	1
11	1	0	0	1	0	1	0	0	0	0
12	1	1	1	0	0	0	0	0	0	0
13	0	0	1	1	0	1	0	0	0	0
14	0	0	0	0	1	0	0	1	1	0
15	0	1	0	0	0	0	1	0	0	1
16	1	0	0	1	0	1	0	0	0	0
17	1	1	1	0	0	0	0	0	0	0
18	0	0	1	1	0	1	0	0	0	0
19	0	0	0	0	1	0	0	1	1	0
20	0	1	0	0	0	0	1	0	0	1

Student

9. **a)** What subjects is student 13 taking? _____

   b) From Table 1, find the *minimum* number of final examination periods that need to be scheduled so that all 10 exams can be administered and no student has a conflict. A conflict is said to occur when a student has two or more final exams scheduled at the same time. _____

More than likely, you used "trial and error" to solve problem 9b). If there were 200 students and 20 subjects, a trial and error solution would be tedious. We now describe a better technique which is based on graph theory.

The solution to the exam scheduling problem is equivalent to finding the *Chromatic Number* of a graph, which is the smallest number of distinct colors required to color the vertices of a graph so that no two adjacent vertices have the same color. As you will see, many practical and important problems are equivalent to finding the chromatic number of a graph.

Consider the graph:

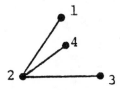

To find the chromatic number, start by "coloring" one vertex, say 2, with a square.

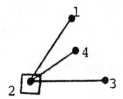

Since vertices 1, 4, and 3 are adjacent to vertex 2, they must be colored with colors other than "square." Further, since vertices 1, 4, and 3 are nonadjacent to each other, all three can be colored with "circle." So we have

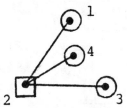

and the chromatic number of the graph is 2.

10. Find the chromatic numbers of the following graphs by "coloring" the vertices with □, ○, △, ◊ or ★.

Chromatic Number

(a) _____

(b) (The Cube) _____

(c) K_4 _____

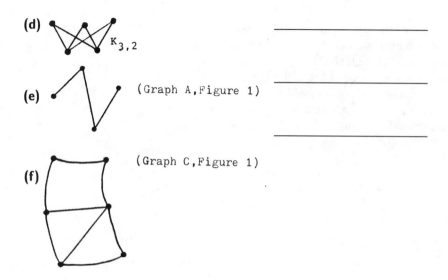

(d) $K_{3,2}$

(e)   (Graph A, Figure 1)

(f)   (Graph C, Figure 1)

11. **(a)** What is the chromatic number of a graph with N vertices and no edges?

_____

(b) What is the chromatic number of a complete graph with N vertices?

_____

(c) What is the chromatic number of $K_{2,2}$? _____

(d) What is the chromatic number of $K_{M,N}$? _____

The exam scheduling problem can be represented by a graph, and the solution to the problem is the chromatic number of that graph. Perhaps the next four questions will suggest how to begin.

12. Under what circumstances would two finals have to be scheduled at different times? _____

13. Under what circumstances could two finals be given at the same time?

_____

14. Do we really need *all* of the information in the table? _____

15. Does it matter that students 1 *and* 6 are taking subjects 1 and 4 or only that *some* student is taking both? _____

Exam scheduling can be portrayed as a graph in which each subject is represented by a vertex. Two vertices are joined by an edge if *any* student is taking both subjects. The interpretation is that if two vertices are joined, then the exams for those two subjects cannot be held at the same time because at least one student would have a conflict. We can assign different colors to such vertices indicating that the corresponding exams must be held at different times.

16. Draw a graph for the exam schedule problem. The vertices are shown in the figure below.

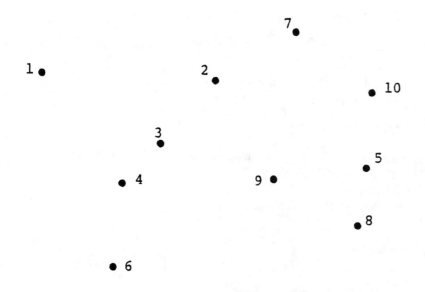

17. Find the chromatic number of the graph (and hence the solution) to the scheduling problem. _____

18. Do any adjacent vertices have the same color? _____

19. Which exams can be scheduled at the same time? _____

20. Is there a coloring which uses only three colors so that no two adjacent vertices have the same color? _____

21. Why or why not? _____ (Hint: Review the "student-subject" matrix.)

A. Chemical Storage Problem

Suppose that 10 chemicals are to be stocked in a warehouse. Some pairs of chemicals are incompatible because mixtures of their fumes are explosive. No such pairs can be stored together. To separate them, we must build partitions and ventilate the rooms with separate exhaust fans; and the more rooms, the greater the cost. Therefore, it is desirable to have only the minimum number of rooms necessary to store all chemicals safely.

All incompatible pairs of chemicals are given below.

Chemical	Incompatible with Chemicals
1	3, 8, 9
2	3, 6, 8, 10
3	1, 2, 4, 6, 7
4	3, 10
5	7, 8, 9
6	2, 3
7	3, 5
8	1, 2, 5
9	1, 5
10	2, 4

22. What is the least number of separate storage areas required? (Hint: Represent the problem as a graph and find the chromatic number. We suggest the reader use a separate sheet of paper for this graph. As in the exam scheduling problem, think carefully about what each vertex and edge in your graph will represent.) _____

23. List the chemicals which can be stored in each area.

Area 1 - _____

Area 2 - _____

Area 3 - _____

Area 4 - _____

Area 5 - _____

Area 6 - _____

24. Is it possible to provide a correct coloring with only 2 colors? Why or why not? _____ (Hint: Consider chemicals 2 ,3, and 6.)

B. Communications Problem

Consider a communications network which transmits 4 bit "letters". The letters are:

A	0000	I	0110
B	0001	J	1010
C	0010	K	1100
D	0100	L	0111
E	1000	M	1101
F	0011	N	1011
G	0101	O	1110
H	1001	P	1111

However, because of errors in transmission, letters which differ in only one bit (or place) are often confused with one another. For example, letters A and B differ only in the 4th bit. But G and J differ in all four bits. The problem is to find the largest set of letters which cannot be confused.

25. Fill in the matrix below. An element will have a "1" if the two corresponding letters can be confused.

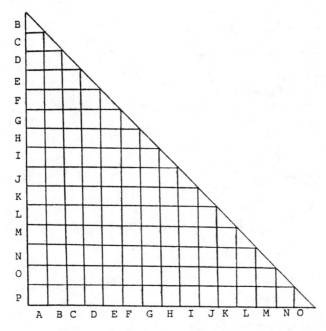

26. **a)** Draw the graph for this problem using the matrix you just developed. The vertices are provided below.

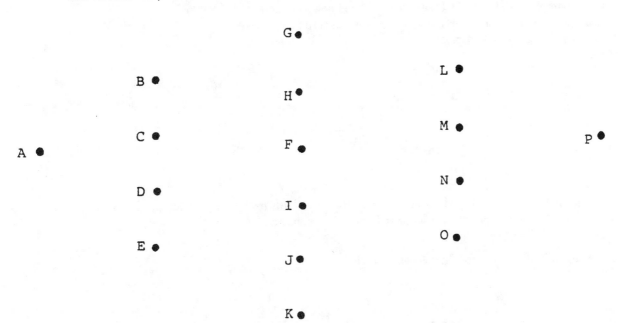

Color the vertices of the graph above. Think about what it means each time a new color is needed. Think about what it means for a set of letters to have the same color.

26. **b)** Find the largest set of letters which can be transmitted and not confused with one another. _____

27. What is the chromatic number of the graph in **26**? _____

28. Is the coloring (and hence the largest set of nonconfusable letters) unique?

_____

The three previous sample problems are small enough that you could draw the graphs and find chromatic numbers by inspection. Larger problems, such as exam scheduling for a university, require a computer. But how can we describe a graph to a computer?

Let's consider the following two graphs.

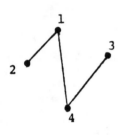 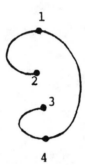

The two look quite different—yet both are *pictures* of the same graph. Both have 4 vertices and edges between vertices 1–2, 1–4, and 3–4. Thus, pictures can be deceiving; the computer descriptions of the two pictures must be identical.

We will define a graph by a matrix. Matrices are useful and powerful tools in the mathematical, physical, and computer sciences and many operations involving them can be easily programmed on most computers. Consider the following "empty" matrix.

Vertex

	1	2	3	4
1				
2				
3				
4				

Vertex

There is a row and a column for each vertex. A "1" is written in the ith row and jth column (called the (i, j)th position) if vertices i and j are connected by an edge. Otherwise, a "0" is placed in that position.

29. Using this rule, fill in the matrix for the following graph.

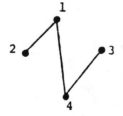

	1	2	3	4
1	___
2	___
3	___
4	___

(Note: such a matrix is called an *adjacency* matrix. Its elements are balanced or symmetrical about a line through the elements (1,1), (2,2), (3,3), and (4,4). Such a matrix is said to be *symmetric* about the main diagonal.)

30. Why does the entry in position (i, j) equal the entry in (j, i)?

Thus, to describe an adjacency matrix completely, you need only give the entries either above or below the main diagonal. (The main diagonal consists of the entries in the positions (1,1), (2,2), (3,3), ...)

31. Fill in the matrices for the graphs below. ((a) is done for you as an example.)

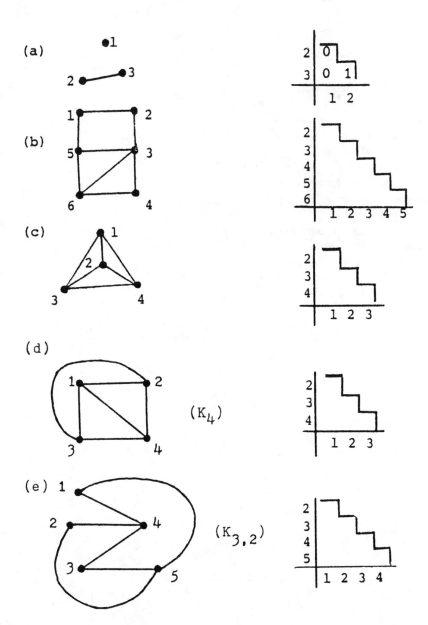

Once the adjacency matrix describing the graph is in the computer, a "coloring algorithm" can be used to solve the problem. (You are asked to invent such an algorithm in the Special Projects section!) As we have seen, vertex coloring is a way of solving many problems in which the presence or absence of some relationship between a pair of vertices can be represented by the presence or absence of an edge between them. We shall return to this topic later in a discussion of the "Four Color Conjecture."

Before we leave the chromatic number, here is another application.

C. Selection of Variables to be Used in Model Building

Analysts often develop mathematical models to describe processes or to make predictions.

The authors have developed such a model, called the Trauma Score, to assess injury severity. The score is based on measurements such as blood pressure and pulmonary rate and can be used to estimate the probability of survival for a patient from measurements at the accident scene or at hospital admission.

In developing the Trauma Score, the authors found by analyzing patient data that 12 variables were powerful indicators of injury severity. An important question was whether all these variables are needed in a predictive model, and if not, which ones to retain. To address this question, they used the correlation coefficient.

In the charts below, each dot represents a different patient; and the x and y dimensions represent blood pressure and pulmonary rate, respectively. The correlation coefficient, denoted r_{xy}, measures the degree of the *linear* relationship among the plotted points.

Consider the following plots.

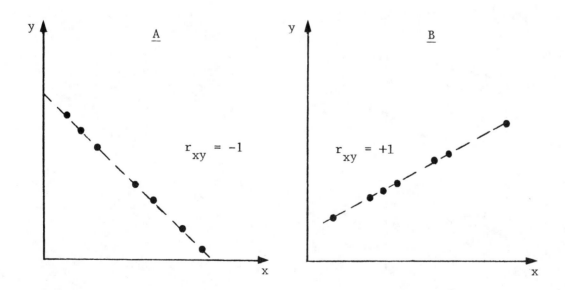

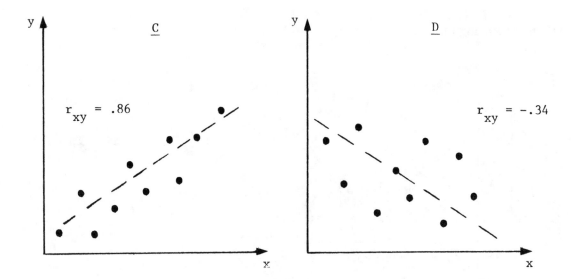

Plots *A* through *D* show the important facts about r_{xy}:

(1)　$-1 \leq r_{xy} \leq 1$;

(2)　$r_{xy} = +1(-1)$ if *all* the (x, y) data points fall precisely on a line with any positive (negative) slope;

(3)　the more the data are scattered, the lower (in absolute value) the correlation; that is, the greater the scatter, the closer r_{xy} is to 0.

Estimate the correlation coefficient for the following data.

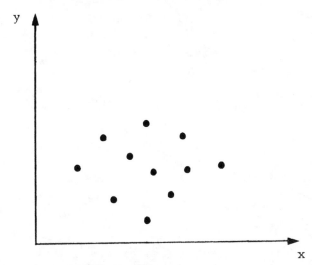

The correlation for these data is nearly zero.

In model building, little is gained by including more than one variable of a highly correlated pair, since when you know one, you essentially know the other. Thus, an analyst may wish to begin with the largest set of candidate variables all pairs of which have "low" correlations.

Table 2 gives the correlations for all pairs of the 12 variables identified in the Trauma Score work as being strongly related to injury severity.

Suppose we agree that two variables have "small correlations" if their correlation coefficient r_{xy} is such that $|r_{xy}| < .60$.

Table 2
Correlations* Among Medical Variables

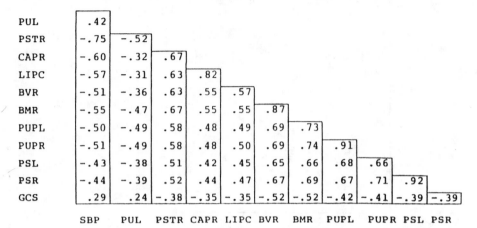

	SBP	PUL	PSTR	CAPR	LIPC	BVR	BMR	PUPL	PUPR	PSL	PSR
PUL	.42										
PSTR	-.75	-.52									
CAPR	-.60	-.32	.67								
LIPC	-.57	-.31	.63	.82							
BVR	-.51	-.36	.63	.55	.57						
BMR	-.55	-.47	.67	.55	.55	.87					
PUPL	-.50	-.49	.58	.48	.49	.69	.73				
PUPR	-.51	-.49	.58	.48	.50	.69	.74	.91			
PSL	-.43	-.38	.51	.42	.45	.65	.66	.68	.66		
PSR	-.44	-.39	.52	.44	.47	.67	.69	.67	.71	.92	
GCS	.29	.24	-.38	-.35	-.35	-.52	-.52	-.42	-.41	-.39	-.39

32. Use the correlations in Table 2, the vertices below, and the concept of coloring to find the largest set of variables in which no pair has a correlation of .60 or greater. _____

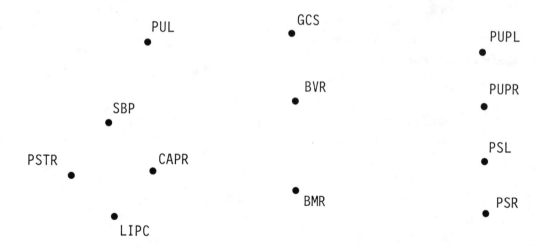

*The correlations were obtained on 2929 trauma patients treated at the Washington Hospital Center, Washington, D. C.

The 12 variables are Systolic Blood Pressure, Pulmonary Rate, Pulmonary Strength, Capillary Refill, Lip Color, Best Verbal Response, Best Motor Response, Pupil Response Left, Pupil Response Right, Pupil Size Left, Pupil Size Right, and Glasgow Coma Scale.

Part II
PLANARITY

In a small community in the Mojave desert, three homes were built each with a well in the back yard.

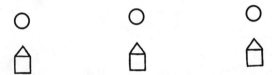

Because of frequent droughts, residents have access to a neighbor's well in addition to their own. However, squabbles broke out among the neighbors. In order to avoid one another, they decided to construct paths from each house to each well so that no two paths crossed.

33. Can you construct edges from each house to each well so that none of the edges intersect? _____

Oh well! If you cannot find a graph in which no two edges intersect, can you find one in which only one intersection occurs? If yes, draw the graph using the houses and wells shown above.

The graph for this problem is a special kind already described in the first section.

34. Give the name for this graph and its shorthand notation. _____

A graph is said to be *planar* if it can be drawn (on a piece of paper—which may be taken to represent a plane) so that its edges intersect only at the vertices.

As examples of planar and nonplanar graphs consider the following:

Graph A:

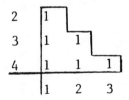

This is a planar graph, as the picture shows.

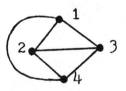

Graph B:

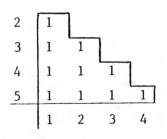

This is not a planar graph. Every picture of graph B has at least two edges which cross.

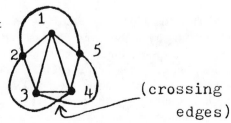

(crossing edges)

To show that a graph is planar, all that is required is one picture in which the edges do not intersect. (Remember, there can be many pictures of the same graph!) But it is difficult to prove that a graph is nonplanar by drawing pictures. Failure to demonstrate planarity pictorially may indicate only a lack of imagination, not nonplanarity.

35. Which of the following pictures are of planar graphs?

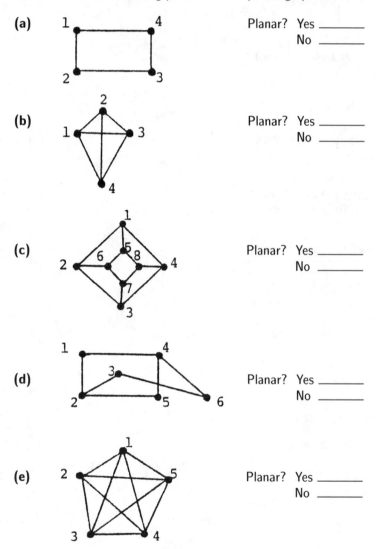

(a) Planar? Yes _____
             No _____

(b) Planar? Yes _____
             No _____

(c) Planar? Yes _____
             No _____

(d) Planar? Yes _____
             No _____

(e) Planar? Yes _____
             No _____

The Wells and Houses Problem is one of the ancient and venerable chestnuts of graph theory, but it's hardly an exciting introduction to the topic of Planarity. Do not be misled. Graph planarity has important modern applications. For example, printed circuit boards (PCB's), which are standard components of such electronic devices as radios, stereos, TVs, and computers, are produced by depositing conductive paths on a sheet of nonconductive material. Such a circuit can be printed on a single sheet only if the graph of the circuit network is planar.

If the circuit can be printed on a single sheet, there are savings in space, weight, and cost.

Because of the complexity of many circuit diagrams, an unsuccessful effort to find a planar representation by trial and error can hardly be considered proof of nonplanarity.

A remarkable theorem proven by K. Kuratowski in 1930, long before PCB's were conceived, establishes criteria for determining whether or not a graph is planar. That theorem states:

A graph is planar if and only if it does *not* contain either of the complete graphs K_5 or $K_{3,3}$ as subgraphs.

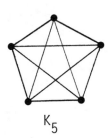

K_5

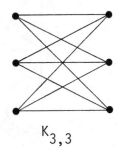

$K_{3,3}$

A graph G' is said to be a *subgraph* of another graph G if all the vertices and edges of G' are also contained in G.

For example, in the figure below, B is a subgraph of A, but C is not.

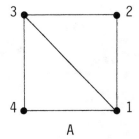

A

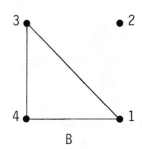

B

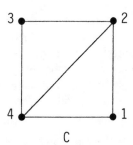

C

Kuratowski's theorem tells us that K_5 and $K_{3,3}$ are the keys to planarity. The reader is already satisfied, by the Wells and Houses Problem, that $K_{3,3}$ is nonplanar; and by Problem 35(e), that K_5 is nonplanar.

Now the reader should consider the following three statements (all true).

(A) If G contains K_M and $M \geq 5$, then G contains K_5.

(B) If G contains $K_{M,N}$ and $M \geq 3$ and $N \geq 3$, then G contains $K_{3,3}$.

(C) If G contains a nonplanar subgraph G', then G is nonplanar.

Kuratowski's Theorem can be divided into two separate statements, as follows:

Kuratowski, Part I: If G contains either K_5 or $K_{3,3}$ as a subgraph, then G is nonplanar.

Kuratowski, Part II: If G is nonplanar, then it contains either K_5 or $K_{3,3}$ as a subgraph.

Part II of this theorem is the really valuable part, as a test for nonplanarity. It says if you can't find K_5 or $K_{3,3}$ in G, then G is planar. But it's the hard part—to prove. Kuratowski's proof came in 1930, nearly 200 years after Euler solved the Königsberg bridge problem.

Consider the following graph which resulted from the correlation matrix, Table 2, and was needed to answer question 32.

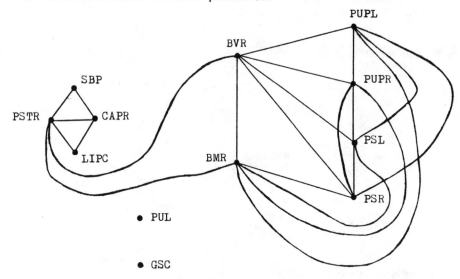

By applying Kuratowski's Theorem, we find that the graph is nonplanar. Note that the subgraph consisting of the vertices BVR, PUPL, PUPR, PSL, and PSR is actually K_5.

Since 1930, several other criteria have been established which guarantee planarity, but none are as simple as Kuratowski's. However, it can be time consuming (even with a computer) to determine the planarity of a large graph by Kuratowski's criteria. One computer algorithm searches for K_5 and $K_{3,3}$ by screening vertices. Only vertices of degree 4 or more are candidates for K_5, and only vertices of degree 3 or more are candidates for $K_{3,3}$.

36. Determine if the following graphs are planar. If planar, provide a "picture" in which no edges intersect. If nonplanar, explain why. _____

Graph	Planar or Nonplanar	Picture or Reason
(a) K_4	_____	
(b) K_6	_____	
(c) K_N; with $N \geq 5$	_____	
(d) The graph whose vertices are the points of a cube and whose edges are the edges of the cube.	_____	

Kuratowski's Theorem does not settle all issues of PCB design and construction. If the theorem shows that the graph representing a circuit is planar, a *picture*, to be used as a model for the PCB, must still be found. If the graph is nonplanar, it is desirable to separate it into the *smallest possible number* of planar graphs. Each of these could then be placed on a single board and the boards connected appropriately. A solution to this problem has been found* and is based upon coloring and chromatic number.

The application of planarity to PCB design and construction is an example of mathematical study preceding its most important uses. However, the concept has been of academic interest for over a hundred years, as explained in the next section.

A. **The Four Color Conjecture**

The graphical concepts of coloring and planarity led to the Four Color Conjecture, one of the most famous mathematical problems ever posed. The conjecture may be stated as follows:

Every planar graph has a chromatic number less than or equal to 4.

The problem arises in the following way. Maps may be considered as planar graphs if vertices are taken as points where borders meet and edges represent borders. As an example, consider the map of South America given below. It translates into the planar graph beside it.

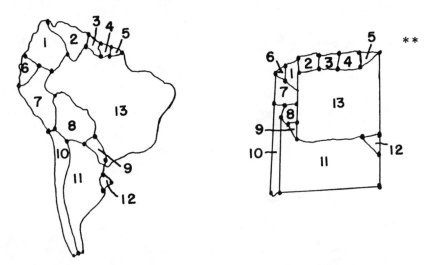

The question is whether 4 colors are *always* sufficient to color the countries so that no two countries which share a border (except perhaps only at a vertex) have the same color. (It can be shown that this statement of the Four Color Problem is equivalent to that given previously.)

---

*On the Planar Decomposition of Interconnection Networks for Computer Aided Electronic Design, Proceedings of 8th Annual Allerton Conference on Circuit and System Theory.

**Figures taken from Reference 8, p. 136.

According to Narsingh Deo (in Reference 9), the four color problem was first proposed by A. F. Möbius in 1840, but did not become well known until 1879 when A. Cayley published it in the Proceedings of the Royal Geographic Society.

For over a century many famous mathematicians tried to prove or disprove the conjecture. In 1976, it was finally shown by two University of Illinois mathematicians, Kenneth Appel and Wolfgang Hanken, that 4 colors are sufficient. A major part of the proof required finding the chromatic number of 1482 graphs, using a computer.

It is worth repeating that this famous problem, apparently belonging to the field of mapmaking, is really a graph theory problem involving the concepts of planarity and chromatic number.

Now on to another simple but useful graph theory concept, that of a tree.

Part III
TREES

A graph is said to be *connected* if we can reach any vertex from any other by traveling along the edges. Graph G of Figure 1 is connected. Graph H is not.

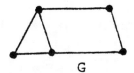

Figure 1

37. Which graphs of Figure 2 are connected?

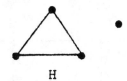

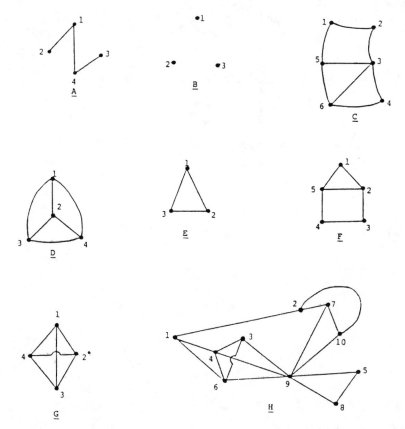

Figure 2

A *cycle* is a path of edges in a graph which starts and ends at the same vertex but which *does not pass through* any edge or vertex more than once.* Note also that a cycle need not include every vertex of the graph. The graph

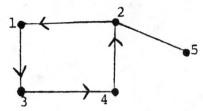

has a cycle, which joins vertices 1, 3, 4, and 2. Note that a cycle can start at any one of those vertices. The graphs

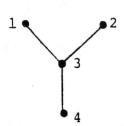   and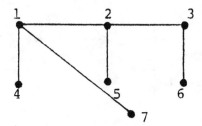

do not contain any cycles.

The concept of a tree is one of the most important in graph theory and has many applications to practical problems. A *tree* is simply a connected graph with no cycles.

38. Which of the graphs below are trees?

      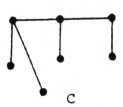

A   B   C

39. Draw trees with 2, 3, 4, 5, and 6 vertices.

_____

*Clearly a cycle would exist between two vertices if there were two distinct edges connecting them.

A. **Applications of Trees**

A telephone company intends to establish communications for the first time in 7 villages (designated A, B, C, D, E, F, G) in an undeveloped country. Each village must be able to communicate with any other village, either directly or through other villages. This communication problem can be represented by the graph in Figure 3.

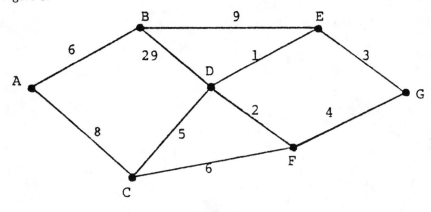

Figure 3

Each vertex represents a village and the numbers on the edges are estimates of the costs (in millions of dollars) of constructing links between villages. Some links are missing because they are prohibitively expensive. The company wants to establish the communications at the smallest construction cost. The graph of Figure 3 is called a *weighted graph* since each edge is associated with a number, called its weight.

Determine your solution to the problem. Hint: Your solution should be a tree containing all of the vertices, but not all of the edges depicted in Figure 3.

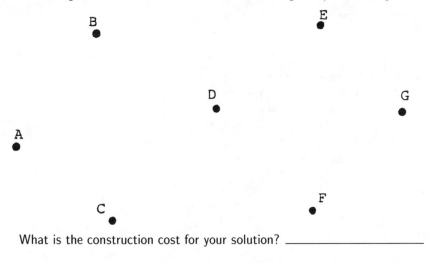

What is the construction cost for your solution? _____

Did you select the edges in the following order?
Edges: DE, DF, EG, DC, AB, AC.

If so, you have invented the Kruskal Algorithm (named for its inventor, J. B. Kruskal) which is: .

Step 1. Choose the edge with least cost.

Step 2. Choose from among the remaining edges, the edge with the least cost which does not produce a cycle.

Step 3. Continue as in Step 2 until all vertices are linked.

The solution using the Kruskal Algorithm is given in the figure below.

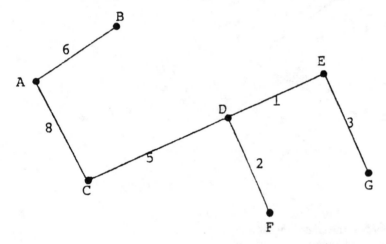

The construction cost is 25 million dollars.

40. What are the paths over which

A and D communicate? _____

B and F communicate? _____

C and G communicate? _____

The graph in the previous figure is a tree. It also happens to be a spanning tree of the graph of Figure 3. A *spanning tree* of a connected graph *G* is a tree in *G* which contains every vertex of *G*. The *weight of a tree* is the sum of the weights of its edges. A *minimal spanning tree* is the spanning tree of minimal weight. Kruskal's Algorithm is one way to find a minimal spanning tree.

Suppose we want to produce the rail network of minimum length which connects 6 cities. The lengths between the city pairs are given below.

	A	B	C	D	E
B	18				
C	9	36			
D	12	14	11		
E	16	9	6	19	
F	23	4	18	26	14

41. Complete the graph which represents this situation.

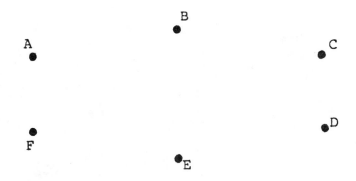

42. Using Kruskal's Algorithm, determine the minimal spanning tree.

        Edge

1.

2.

3.

4.

5.

Another algorithm for the minimal spanning tree is R. C. Prim's, which proceeds as follows. First, fill in the shortest edge, bringing its two nodes into the tree; second, fill in the shortest remaining edge having one node inside the tree and one outside, thus bringing in another node; and finally, repeat the second step until all the outside nodes are brought in.

B. **Application of Minimal Spanning Trees to "Clustering"**

Clustering is a valuable tool in exploratory data analysis. It is particularly popular in medicine, biology, archeology, and economics. Clustering methods separate data into subsets, so that two data points from the same cluster are considered similar, and two data points from different clusters are considered dissimilar. Over the last twenty years, many clustering approaches have been devised. In this section, we present an interesting application of minimal spanning trees to clustering.

Table 3 contains intensive care unit data from 35 patients who died from serious injuries. Each patient is characterized by four numbers, which reflect the poorest conditions experienced by the respiratory, kidney, liver, and central nervous systems during the intensive care period. Larger numbers reflect poorer patient condition.

Table 3

Intensive Care Unit Data from 35 Patients
Who Died Following Serious Injuries

Patient	Maximum Respiratory Index	Maximum Kidney Index	Maximum Liver Index	Maximum Central Nervous System Index
1	1	2	2	12
2	5	12	54	8
3	3	4	13	12
4	3	8	21	7
5	4	1	0	9
6	2	2	4	12
7	5	8	14	10
8	4	3	1	12
9	3	2	3	12
10	3	1	1	12
11	2	1	2	12
12	8	2	1	12
13	11	8	15	12
14	11	6	18	12
15	10	10	3	12
16	7	3	8	12
17	7	4	1	12
18	4	2	2	12
19	4	3	1	12
20	10	4	13	9
21	2	2	1	4
22	2	1	1	12
23	3	2	1	12
24	6	2	1	12
25	4	2	2	12
26	5	3	3	12
27	3	2	3	12
28	6	2	4	12
29	2	2	1	9
30	3	2	1	11
31	2	2	1	12
32	6	9	56	12
33	2	3	1	12
34	2	3	1	11
35	1	1	2	12

Represent each patient as a 4-dimensional vector, $\mathbf{X} = (x_1, x_2, x_3, x_4)$, where

x_1 = maximum respiratory index;
x_2 = maximum kidney index;
x_3 = maximum liver index; and
x_4 = maximum central nervous system index.

43. What are the x_1, x_2, x_3, x_4 values for

Patient 7 _____,

Patient 15 _____,

Patient 23 _____.

It is useful to think of each vector $\mathbf{X}$ as a point in 4-dimensional space. In our example, we have 35 points, one for each patient. The city block distance, D, between points $\mathbf{X} = (x_1, x_2, x_3, x_4)$ and $\mathbf{Y} = (y_1, y_2, y_3, y_4)$ is defined as $D = |x_1 - y_1| + |x_2 - y_2| + |x_3 - y_3| + |x_4 - y_4|$.

The city block distance between the points for patients 8 and 16 is $|4 - 7| + |3 - 3| + |1 - 8| + |12 - 12|$ or 10.

44. Compute D for

Patients 7 and 12 _____,

Patients 9 and 34 _____,

Patients 10 and 25 _____.

Table 4 is a city block distance matrix for the 35 patients. The number at the intersection of row 7 and column 22 is the distance between the points representing patients 7 and 22. The number at the intersection of row i and column j is the distance between the points representing patient i and patient j.

45. From the distance matrix, determine the distance between the points representing:

Patient 4 and Patient 23 _____,

Patient 13 and Patient 35 _____,

Patient 10 and Patient 20 _____.

To cluster the 35 points in 4-dimensional space, we first construct a minimal spanning tree, thinking of the points as vertices of a graph and the distance between points as weighted edges.

We begin the construction of the minimum spanning tree by noting there are three zeros in Table 4.

46. What do these zeros mean? _____
_____

Figure 4 contains the vertices of the graph and the first 9 edges of the minimal spanning tree; namely, those edges with city block distance 1. The 9 edges, obtained by scanning the rows of Table 4 starting with the first row, are:

1, 35	22, 31
10, 22	23, 30
10, 23	31, 33
11, 22	33, 34
11, 35	

The edge (23, 31) is not listed because it creates a cycle.

Table 4

	1	2	3	4	5	6	7	8	9	10	11	12	13	14	15	16	17	18	19	20	21	22	23	24	25	26	27	28	29	30	31	32	33	34	35
1		70	15	32	9	3	24	5	3	4	2	8	29	30	18	13	9	3	5	25	10	3	3	6	3	6	6	7	5	4	2	66	3	4	1
2			55	40	67	67	46	67	67	70	70	70	53	52	62	61	67	67	67	49	70	71	69	68	67	63	67	65	67	68	70	10	69	68	71
3				17	20	12	9	14	12	15	15	19	14	15	23	10	16	14	14	10	23	16	14	17	14	13	12	14	18	15	15	51	14	15	16
4					31	29	12	31	29	32	32	36	19	18	32	27	33	31	31	21	30	33	31	34	31	30	29	31	29	30	32	44	31	30	33
5						10	23	6	8	5	7	9	32	43	21	16	10	6	6	22	9	6	6	7	6	8	8	10	4	5	7	69	8	7	8
6							21	6	2	5	3	9	26	27	17	10	10	4	6	22	11	4	4	7	4	5	2	4	6	5	3	63	4	5	4
7								21	21	24	24	24	9	14	20	15	21	21	21	11	28	25	23	22	21	18	21	19	23	22	24	46	23	22	25
8									4	3	5	5	26	24	15	10	4	2	0	22	11	4	2	3	2	3	4	6	6	3	3	63	2	3	6
9										3	3	7	26	27	15	10	8	2	4	22	11	4	2	5	2	3	0	4	6	3	3	63	4	5	4
10											2	6	29	30	18	13	7	3	3	25	10	1	1	4	3	6	3	7	5	2	2	66	3	4	3
11												8	29	30	18	13	9	3	5	25	10	1	3	6	3	6	3	7	5	4	2	66	3	4	1
12													23	24	12	9	3	5	5	19	14	7	5	2	5	6	7	5	9	6	6	64	7	8	9
13														5	15	16	22	26	26	10	37	30	28	25	26	23	26	22	32	29	29	47	28	29	30
14															20	17	23	27	27	11	38	31	29	26	27	24	27	23	33	30	30	46	29	30	31
15																15	11	15	15	19	26	19	17	14	15	12	15	13	21	18	18	58	17	18	19
16																	8	10	10	12	21	14	12	9	10	7	10	6	16	13	13	55	12	13	14
17																		6	4	18	15	8	6	3	6	5	8	6	10	7	7	61	6	7	10
18																			2	22	11	4	2	3	0	3	2	4	6	3	3	63	4	5	4
19																				22	11	4	2	3	2	3	4	6	6	3	3	63	2	3	6
20																					27	26	24	21	22	19	22	18	23	23	25	55	24	23	26
21																						9	9	12	11	14	11	15	5	8	8	74	9	8	11
22																							2	5	4	7	4	8	4	3	1	67	2	1	2
23																								3	2	5	2	6	4	1	1	65	2	3	4
24																									3	4	5	3	7	4	4	62	5	6	7
25																										3	2	4	6	3	3	63	4	5	4
26																											3	3	9	6	6	60	5	6	7
27																												4	6	3	3	63	4	5	4
28																													10	7	7	59	8	9	8
29																														3	3	69	4	3	5
30																															2	66	3	2	5
31																																66	1	2	3
32																																	65	66	67
33																																		1	4
34																																			5
35																																			

Continue the construction of the minimum spanning tree, on Figure 4, by adding edges of city block distance 2 to the graph. Do not include edges that create cycles.

You should have added edges:

6, 9

8, 18

8, 23

9, 18

12, 24

47. Do you have a minimum spanning tree yet? _____

Add the edges of city block distance 3. They are:

8, 24

8, 26

12, 17

24, 28

48. Do you have a minimum spanning tree yet? _____

The edge (5, 29) is the only edge of distance 4 that does not create a cycle. Add it to the graph.

The edges (13, 14) and (21, 29) are the only edges of distance 5 that do not create cycles. Add them to the graph.

The edges (5, 8) and (16, 28) are the only edges of distance 6 that do not create cycles. Add them to the graph.

All edges of distance 7 and distance 8 create cycles.

You can complete the minimum spanning tree by adding:

> edges (3, 7) and (7, 13) of distance 9;
> edges (2, 32), (3, 16), and (3, 20) of distance 10;
> edge (15, 17) of distance 11;
> edge (4, 7) of distance 12; and
> edge (2, 4) of distance 40.

Figure 5 on page 33 contains a complete minimal spanning tree.

49. Does it agree with yours? _____

By definition, the minimal spanning tree connects *all* vertices. It does not define clusters. But it does contain information useful for clustering. For example, *an edge with a large weight (distance) relative to the other edges emanating from its vertices can be "cut" to form two clusters.* Consider vertex (patient) 23. It is connected to vertices 10, 30, and (8, 19). The distance from 23 to (8, 19) is 2, twice the distance to the other vertices. If the (8, 23) edge is cut, a

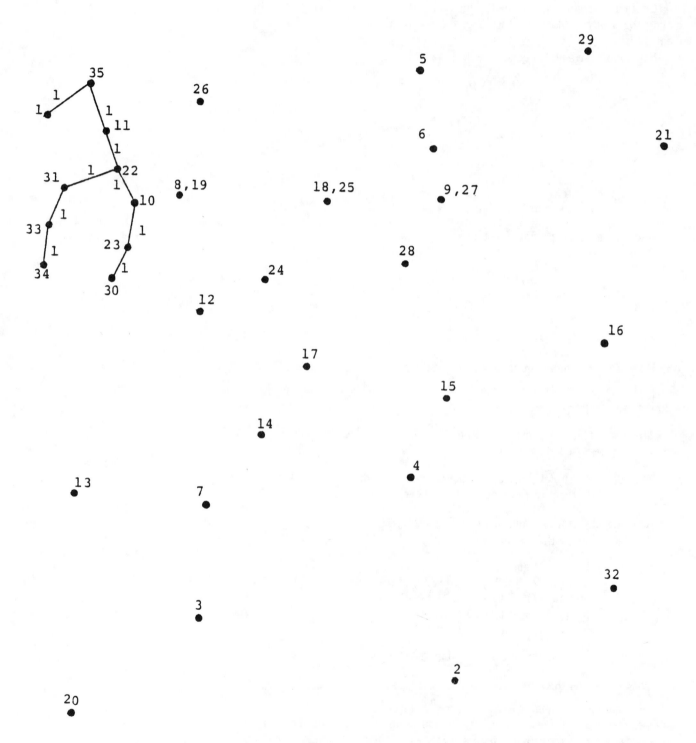

Figure 4

VERTICES OF GRAPH MODEL OF 35 PATIENTS
AND FIRST 9 EDGES OF A MINIMAL SPANNING TREE

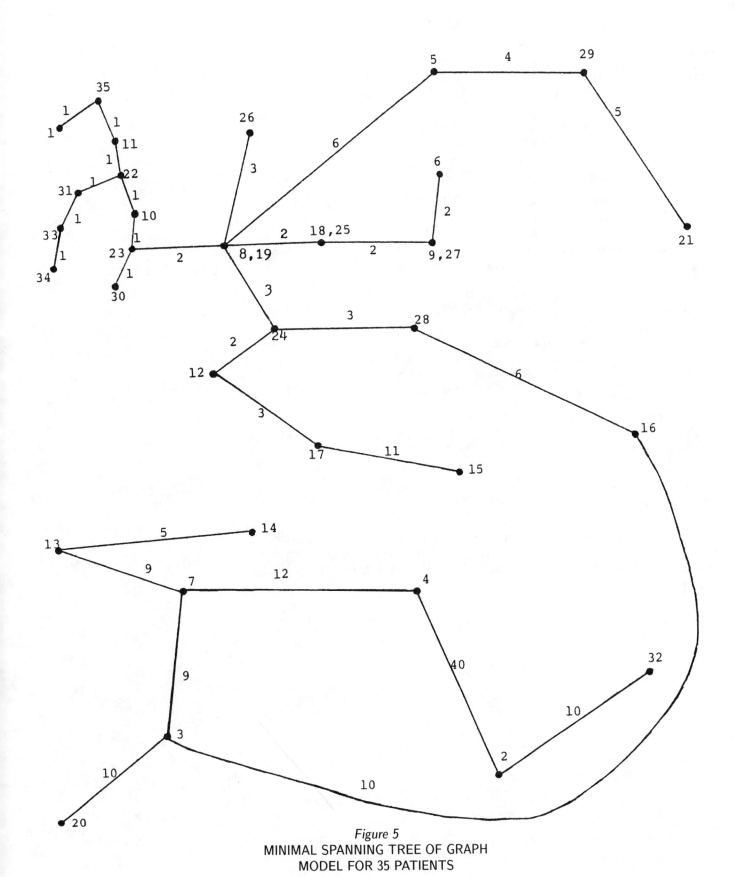

Figure 5
MINIMAL SPANNING TREE OF GRAPH
MODEL FOR 35 PATIENTS

cluster is formed of patients [1, 10, 11, 22, 23, 30, 31, 33, 34, 35]. In Figure 5 if we cut edges (8, 23), (5, 8), (15, 17), (16, 28), (2, 4) we get the clusters:

Cluster A: [1, 10, 11, 22, 23, 30, 31, 33, 34, 35]
Cluster B: [6, 8, 9, 12, 17, 18, 19, 24, 25, 26, 27, 28]
Cluster C: [5, 21, 29]
Cluster D: [15]
Cluster E: [3, 4, 7, 13, 14, 16, 20]
Cluster F: [2, 32]

Hence this set of nonsurvivors can be characterized as belonging to 6 categories (clusters) which have reasonable medical interpretations.

Cluster

A	Grave Central Nervous System (CNS) Problems
B	Grave CNS and Moderate Respiratory Problems
C	Serious CNS Problems
D	Grave CNS, Respiratory, and Renal Problems
E	Grave problems with all 4 systems or of varied nature
F	Grave Hepatic and CNS Problems

As mentioned previously, there are many clustering methods. In the Special Questions/Projects Section, you will have an opportunity to invent your own clustering techniques.

C. A Shortest Path Problem

The notion of a tree also provides a model for another interesting problem. Consider a graph G, whose edges are labeled with the "cost" in time or dollars of traveling between the related vertices, which may represent cities. An example is given below.

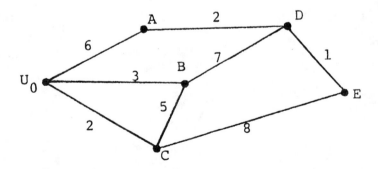

The question naturally arises: How can one get from the starting point, U_0, to another vertex in the graph for the least cost?

50. For the graph above list, by inspection and in order, the vertices encountered and the minimal cost to get from U_0 to any other city.

Travel from U_0 to	Vertices Encountered	Minimal Cost of Path
A		
B		
C		
D		
E		

51. Were all the edges used in finding the shortest paths? _____

_____

52. Which edges were excluded from all paths? _____

53. Place on the graph below, only those edges which were part of the shortest path from U_0 to another vertex.

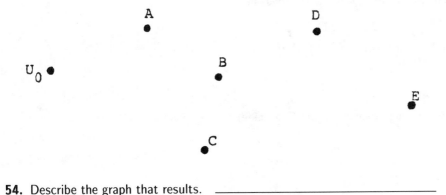

54. Describe the graph that results. _____

_____

_____

Deriving the shortest path by inspection is easy for graphs with few vertices. For larger problems, it helps to have a systematic approach (algorithm). Such an algorithm is described now.

Each vertex of the graph G will be assigned to one of two subsets* S and $\overline{S}$. At first S contains U_0 only. All other vertices are in $\overline{S}$. At each step in the algorithm identify *the* vertex, V, in $\overline{S}$ which can be reached by the shortest path from U_0. That path must pass through only vertices already in S and the single edge linking V with those vertices. Vertex V is then moved from $\overline{S}$ to S, and the process is repeated.

To illustrate the algorithm, a graph is shown on the next page.

Dotted lines are used for edges. These lines are made solid as vertices are added to S.

Algorithm: At start, $S = \{U_0\}$, $\overline{S} = \{A, B, C, D, E\}$

_____

*A set X is said to be a *subset* of the set Y if every element of X is also an element of Y. For example, if $X = \{1, 2, 3, 4, 5\}$ and $Y = \{1, 3\}$, then Y is a subset of X.

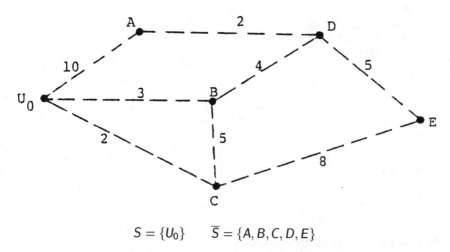

$$S = \{U_0\} \qquad \overline{S} = \{A, B, C, D, E\}$$

Stage 1. Add to S the vertex in $\overline{S}$ which is closest to U_0. At this point, the only choices are those which are adjacent to U_0, which leads to the following table:

	Vertex in $\overline{S}$	Min Dis from U_0 on Path thru Adjacent Vertex	Vertex Path
Adjacent	A	10	U_0A
to U_0	B	3	U_0B
	C	2	U_0C

Since the distance is smallest from U_0 to C, the edge joining U_0 and C is made solid and vertex C is added to S.

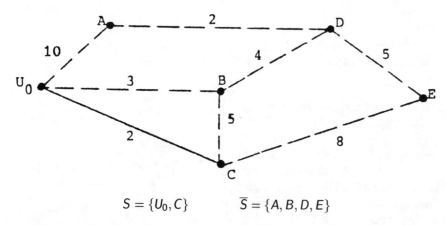

$$S = \{U_0, C\} \qquad \overline{S} = \{A, B, D, E\}$$

Stage 2. Now we add to S the vertex in $\overline{S}$ which is adjacent to U_0, C, or both, which gives the shortest path from U_0 to any vertex in $\overline{S}$. The following table lists the possibilities.

	Vertex in S	Min Dis from U_0 on Path thru Adjacent Vertex	Vertex Path
Adjacent	A	10	U_0A
to U_0	B	3	U_0B
Adjacent	B	7	U_0CB
to C	E	10	U_0CE

Since the minimum distance is associated with vertex B, when considered as adjacent to U_0, B is added to S and the U_0B edge is made solid.

Continue the process until $S = \{U_0, A, B, C, D, E\}$. At that point, there will be a single path from U_0 to any other vertex in the graph. That path will be the one with the least possible cost. The resulting graph is:

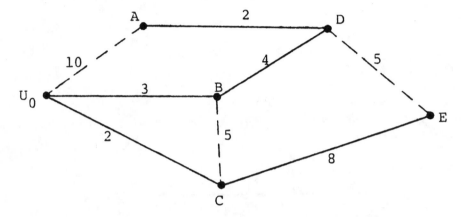

This algorithm, which was invented by E. W. Dijkstra, has been called a "tree growing" algorithm since, at any stage, the solid edges form a tree connecting the vertices thus far included in S.

Use Dijkstra's algorithm to find the shortest path from U_0 to the other vertices of the following graph. Make the edges solid as vertices are added to S.

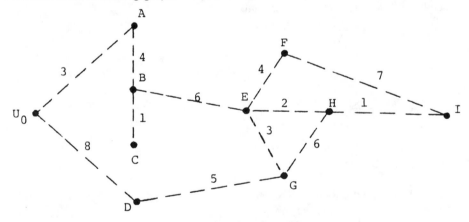

55. List the path and the minimum distance from U_0 to every vertex.

Vertex	Path from U_0	Distance
A	U_0	
B	U_0	
C	U_0	
D	U_0	
E	U_0	
F	U_0	
G	U_0	
H	U_0	
I	U_0	

Part IV
DIRECTED GRAPHS*

In all the material and applications discussed so far, a graph G was composed of a set of vertices and a set of edges which connect some or all pairs of vertices. However, a slight change in definition opens up another area of graph theory which has important applications. In directed graphs, an edge, which as before connects two vertices, has direction. Consider the two graphs below:

Both graphs consist of two vertices and one edge. But the edge directions are different and so the graphs are distinct *directed graphs*, or di-graphs.

We may think of edge direction as indicating a flow of information, material, knowledge, or electricity between vertices. To represent a two way flow between vertices, two edges are needed, one in each direction.

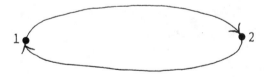

Both matrix representation and the concept of degree undergo changes in their application to di-graphs. In undirected graphs, the degree of a vertex is the number of edges "attached" to it. Since edges in di-graphs have direction, the concept of degree is expanded. The *in-degree* of a vertex in a di-graph is the number of edges coming *into* that vertex from all other vertices. The *out-degree* of a vertex is the number of edges *from* that vertex to all others.

Recall that in the matrix representation of a graph, only the portion above or below the diagonal had to be specified to define the graph completely. This is *not* the case with di-graphs. For example, consider the following di-graph, and

*Our study of directed graphs requires some knowledge of the matrix operations of addition and multiplication. Students not familiar with these operations should proceed directly to Appendix III.

its adjacency matrix.

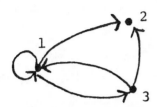

		To		
		1	2	3
	1	1	1	1
From	2	0	0	0
	3	1	1	0

Note that vertex 1 has a directed edge, or loop, to itself. This is included in the matrix as a "1" in the position corresponding to matrix position (1,1).

56. What happens to any "flow" to vertex 2? _____

_____

Such a vertex is called a *sink* and has all zeros or at most a single, special "1", in the From row of its matrix representation.

57. **a)** What is the single admissible "1" in the "From" row of a sink?

_____

b) What is the maximum possible out-degree of a sink?

_____

c) What is the in-degree of vertex 3 in the sample graph above?

_____

58. Fill in the adjacency matrices of the following di-graphs.

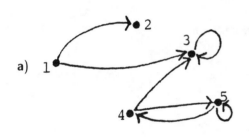

a)

		To				
		1	2	3	4	5
	1					
	2					
From	3					
	4					
	5					

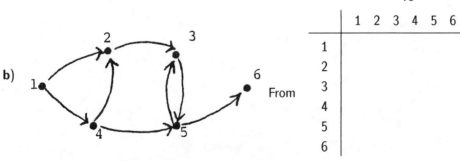

b)

		To					
		1	2	3	4	5	6
	1						
	2						
From	3						
	4						
	5						
	6						

59. Derive di-graphs which represent the following situations:

   a) *University Hierarchy*—Is headed by a President, who has reporting to him the Deans of Admission, Finance, and Academics. Reporting to the Dean of Academics are the Head of the College of Arts and Science and the Head of the College of Business Administration. Reporting to the Head of the College of Arts and Science are the Chairpersons of the Mathematics, Statistics, Biology, Psychology, and Physics Departments. Individual professors report to each Chairperson.

b) *Family Tree*—Starting with your four grandparents, derive a di-graph of your family tree ending with you.

There are many other relationships and activities (flowcharts, tournaments, rumor networks, ...) which di-graphs can represent. The following sections discuss Critical Paths and Reachability, two di-graph concepts with important applications.

A. **Critical Paths**

Suppose, as do the authors, you enjoy breakfast. However, as you have a busy day, you want to prepare breakfast without wasting time. Suppose that the following activities are to be performed.

Activity	Abbreviation	Time
Set Table	ST	2
Fry Sausage	FS	13
Make Coffee	C	4
Make Toast	T	3
Fry Eggs	FE	4
Eat	E	9

Of course, these activities could be performed sequentially in 35 minutes, and represented with the di-graph:

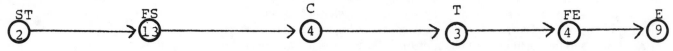

The vertices are labeled with activity abbreviations and the times required to perform the activity are written inside the vertices.

There is a way to reach E more quickly. Consider the order in which the activities must occur.

(1) Since there is only one frying pan, it is necessary (and desirable) to fry the sausage before the eggs;

(2) No particular order is required among the other activities, of setting the table, making coffee and toast, or between them and the two activities requiring the frying pan;

(3) All other activites must be accomplished before eating.

Such information about the activities can lead to a di-graph of breakfast preparation.

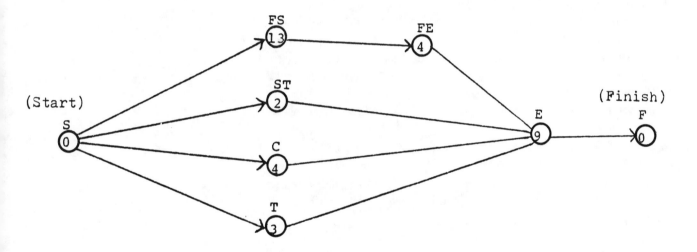

Two nodes, S and F (for start and finish), have been added (for convenience of discussion) and have 0 times. Such a di-graph is called an activity network.

Convince yourself that the di-graph represents the activities and the time-phasing as described in 1, 2, and 3 above.

60. With the di-graph, find the shortest time in which breakfast can be prepared and eaten. _____

61. There are four paths from S to F in the breakfast di-graph. Fill in the table for the three remaining paths.

Path Number	Activity Sequence	Time from S to F
1	S – T – E – F	0 + 3 + 9 + 0 = 12
2	_____	_____
3	_____	_____
4	_____	_____

The path with the longest time from S to F is S – FS – FE – E – F, which requires 26 minutes.

The shortest time to complete all activities is exactly the time for the longest path from S to F. *The path with the longest time is called the Critical Path.* *

Therefore, in order to have breakfast as quickly as possible we must make sure that the table is set (ST), and the toast (T) and coffee (C) are made *while* the sausage and eggs are frying (FS and FE).

There is another important point regarding the activities along the critical path. If any activity *on the critical path* is started later than the earliest possible time, the time required for the whole process increases by exactly that amount.

For example, suppose the cook first set the table (ST) and then started to fry the sausage (FS). The 2 minute table setting time would add to the shortest total time to cook and eat the meal, which would then be 28 minutes. Thus, in order to be time-efficient, a cook must carefully "manage" the times at which activities FS and FE are started or else the project will take longer than necessary.

Some leeway, or slack, is possible in the starting times for activities not on the critical path. For example, it is not necessary (or desirable, in this case) to make the toast, T, as soon as the sausage is started. But how much slack time is possible while still completing the project in the shortest possible time?

To answer that question, two definitions are needed and apply to every activity.

Earliest Time: The earliest time at which an activity can start if all activities which must precede it are started as early as possible;

Latest Time: The latest time an activity can start while not delaying completion of the entire process beyond its earliest completion time.

The Slack Time for an activity is defined as

$$SLACK\ TIME = LATEST\ TIME - EARLIEST\ TIME$$

and is the length of the time interval in which the activity must start to not delay the completion of the entire process. Consider the following table.

Event	Earliest Time	Latest Time	Slack Time
FS	0	0	0
ST	0	15	15
C			
T			
FE			
E			

---

*Think about it! Every activity must be performed. Every activity is at least on one path. Activities on a path have an order in which they must be performed. The minimum time needed to complete all paths is at least as large as the time required for the longest path. But other paths can be completed in no more time than the longest path. Thus, the minimum time required for all paths is at most the time required of the longest path. Therefore, the minimum time required to complete all paths equals the time required for the longest path.

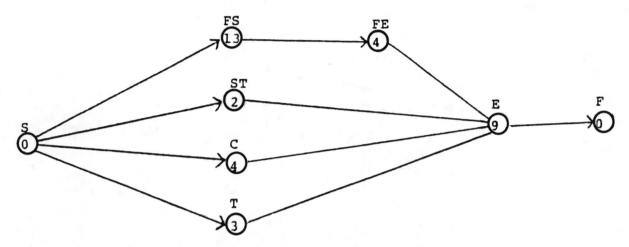

The entries for FS may be explained as follows. (You may find it helpful to refer to the di-graph above.) Since no event precedes FS, the earliest time it can start is 0. If the process is to be completed in the minimum time of 26 minutes, then since FE and E follow FS and require 13 minutes and FS itself requires 13 minutes, FS can start no later than time = 0. Hence, the slack time for FS is zero.

For ST, the earliest time is 0, since no activity precedes it. If the process is to be completed in 26 minutes, ST must start no later than 11 minutes before the 26 minute mark (15 minutes after start time). This is because both activities ST and E must be accomplished and they require a total of 11 minutes. Thus, ST can be started at any time between 0 and 15 minutes into the process and so ST is said to have a slack time of 15.

62. Fill in the previous table for activities C, T, FE, and E.

63. What do you notice about events which have zero slack time?

_____

_____

The critical path method was developed in the late 1950's to aid in the development of several complex defense systems, like the Polaris submarine. The method was said to enhance coordination among thousands of contractors and many government agencies, and shortened project completion time by more than two years. It highlights the activities which are critical to insure the earliest completion and allows managers to concentrate on keeping those activities on schedule. Activities with larger slack time require less management attention.

In practical applications of the critical path method, flow diagrams can be extremely complex. For example, the figure on the next page was reproduced from reference 6 and shows the System Flow Plan for the propulsion (engine) component of the Polaris submarine. The critical path is highlighted. Clearly, computer processing of such networks is required.

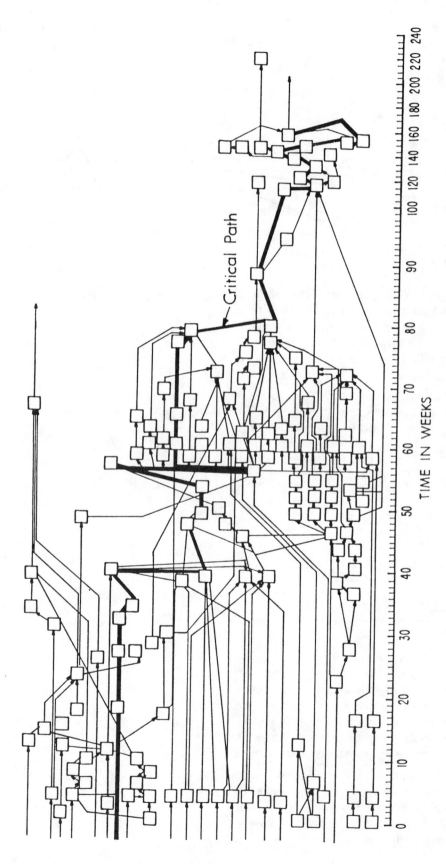

Critical Path

TIME IN WEEKS

System Flow Plan – Propulsion Component.

Other critical path applications include the scheduling and management of building projects, research projects, and virtually any project in which the completion of many activities is required. In addition, other capabilities have been added to critical path analyses which are useful in identifying least cost or least risk paths. The interested reader is encouraged to consult reference 7 for a thorough review of network analysis techniques.

Suppose that activities A_1 through A_7 are *all* required in a certain manufacturing process. The activities, the time each requires and the activities which must immediately precede it are given below.

Activity	Time Required	Immediately Preceding Activities
A_1	10	None
A_2	5	None
A_3	9	None
A_4	18	$A_1; A_2$
A_5	6	A_2
A_6	2	$A_3; A_4; A_5$
A_7	8	A_4

64. With this information, draw the di-graph which represents the situation.

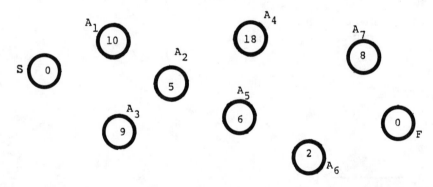

65. How many paths are there from S to F? _____

66. Fill in the table below for each path from S to F for the di-graph above.

Path Number	Activity Sequence	Time from S to F
1	_____	_____
2	_____	_____
3	_____	_____
4	_____	_____
5	_____	_____
6	_____	_____

67. What is the minimum time needed to complete the process?

_____

68. List the activities on the critical path.

_____

You will be asked to fill in the table below, but let's review a few points first. In particular, the concept of latest time may have caused some confusion.

The latest time for an event *not* on the critical path is the latest time that the event can begin and not cause the project time to exceed that of the critical path. Consider for example event A_2. It must be completed before event A_4 (which is on the critical path) can begin. But A_4 cannot begin earlier that $T = 10$, since that is the time required to complete A_1 which must also precede A_4. Thus, since the time to complete A_2 is 5 units, it can start as late as $T = 5$ and still be completed by $T = 10$. Therefore, the latest time for A_2 is 5.

Events lying on the critical path have slack time equal to zero—hence the latest time must equal the earliest time.

Event	Earliest Time	Latest Time	Slack Time
A_1			
A_2	0	5	5
A_3			
A_4	10	10	0
A_5			
A_6			
A_7			

69. Complete the table above.

70. List, in descending order of importance, the attention which you as the manager of this project would give to the different activities, A_1 through A_7.

_____

_____

Consider the following di-graph.

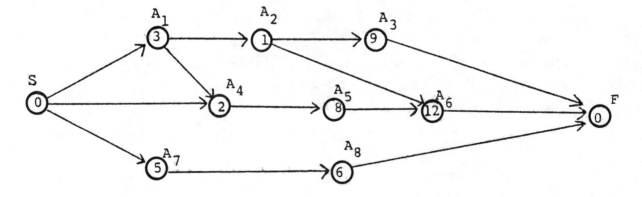

71. Fill in the "precedence table" for this di-graph.

Activity	Immediately Preceding Activities
A_1	
A_2	
A_3	
A_4	
A_5	
A_6	
A_7	
A_8	

72. Fill in the table below, which describes the paths for this di-graph. (There may not be 8 distinct paths.) Identify with an * the critical path.

Path Number	Activity Sequence	Time from S to F
1	_____	_____
2	_____	_____
3	_____	_____
4	_____	_____
5	_____	_____
6	_____	_____
7	_____	_____
8	_____	_____

73. Complete the following table.

Event	Earliest Time	Latest Time	Slack Time	Management Priority
A_1				
A_2				
A_3				
A_4				
A_5				
A_6				
A_7				
A_8				

There are many aspects of critical path methods which have not been considered here. The interested reader should consult the references for more information.

B. **Reachability**

Directed graphs and their applications to modeling real world situations lead to another interesting concept known as reachability.

To illustrate reachability, suppose a directed communications network exists among a set of cities. (In this problem, two-way communication is not always possible because of transmitter and receiver difficulties between cities.) The di-graph for the situation is:

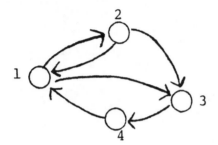

An interesting question which arises in situations like this is: Is it possible to get a message from each vertex to any other? That is, is every vertex "reachable" from every other? We shall see that the answer is obtainable by manipulating the adjacency matrix of the di-graph and then we shall explore some additional applications of reachability.

74. Fill in the adjacency matrix, M, below, for the communication problem.

		To			
		1	2	3	4
From	1				
	2				
	3				
	4				

Let's review the definition of M's elements. The element in the ith row and jth column is "1" if there is a directed edge going *from* the ith vertex *to* the jth vertex. Otherwise, the entry is a zero. Interpreted another way, that element is "1" if there is a *one*-edged path *from* vertex i to vertex j.

Consider what happens when the adjacency matrix is multiplied by itself.

From the rules for matrix multiplication, the (i, j)th entry of the resulting matrix, M^2, is formed by the *inner product* of the ith row and jth column of the original matrix, M. For example, in the diagram below, the element of M^2, in the 1st row and 4th column is found by multiplying each entry of the 1st row by the corresponding entry in the 4th column of M and adding the products. The result is $M^2(1, 4) = 1$.

$$
\begin{array}{ccc}
M & M & M^2 \\[4pt]
\begin{bmatrix} 0 & 1 & 1 & 0 \\ 1 & 0 & 1 & 0 \\ 0 & 0 & 0 & 1 \\ 1 & 0 & 0 & 0 \end{bmatrix}
\cdot
\begin{bmatrix} 0 & 1 & 1 & 0 \\ 1 & 0 & 1 & 0 \\ 0 & 0 & 0 & 1 \\ 1 & 0 & 0 & 0 \end{bmatrix}
=
\begin{bmatrix} & & & ① \end{bmatrix}
\end{array}
$$

But what is the meaning of this "1" entry? Note that the 1 is the product of $M(1,3)$ and $M(3,4)$. Since $M(1,3)$ and $M(3,4)$ are both "1", there are directed edges from vertex 1 to vertex 3 and from vertex 3 to vertex 4. Hence, there is a 2-edged path from vertex 1 to vertex 4, which passes through vertex 3. We can observe that that combination gives rise to the only nonzero term in the sum which constitutes $M^2(1,4)$. We can conclude that there is a single 2-edged path from vertex 1 to vertex 4.

In general, the (i,j)th entry of the matrix, M^N, is exactly the number of N-edged paths which start at the jth vertex and go to the ith vertex.

75. Compute the M^2 matrix for our communications example.

$$\begin{bmatrix} 0 & 1 & 1 & 0 \\ 1 & 0 & 1 & 0 \\ 0 & 0 & 0 & 1 \\ 1 & 0 & 0 & 0 \end{bmatrix} \cdot \begin{bmatrix} 0 & 1 & 1 & 0 \\ 1 & 0 & 1 & 0 \\ 0 & 0 & 0 & 1 \\ 1 & 0 & 0 & 0 \end{bmatrix} = \begin{bmatrix} \_ & \_ & \_ & 1 \\ \_ & \_ & \_ & \_ \\ \_ & \_ & \_ & \_ \\ \_ & \_ & \_ & \_ \end{bmatrix}$$

76. How many two-edged paths exist:

   a) from vertex 1 to vertex 3? _____

   b) from vertex 2 to vertex 4? _____

Thus, to find the total number of paths, of any length, which connect two vertices, it seems that we will have to continue to compute $M^2, M^3, M^4, M^5, M^6, \ldots$ forever. Not so, fortunately! It can be shown that for di-graphs (or graphs) with K vertices, if two distinct vertices can be joined by a path, they can be joined by one which contains $K - 1$ or fewer edges. Therefore, for our communications problem we only need to compute up to M^3 since we have only 4 vertices.

77. Compute M^3 for our communications problem. $[M \times M^2 = M^3]$ (See footnote.)

$$\begin{bmatrix} 0 & 1 & 1 & 0 \\ 1 & 0 & 1 & 0 \\ 0 & 0 & 0 & 1 \\ 1 & 0 & 0 & 0 \end{bmatrix} \cdot \begin{bmatrix} 1 & 0 & 1 & 1 \\ 0 & 1 & 1 & 1 \\ 1 & 0 & 0 & 0 \\ 0 & 1 & 1 & 0 \end{bmatrix} = \begin{bmatrix} & & & \\ & & & \\ & & & \\ & & & \end{bmatrix}$$

78. **a)** Interpret the (3,4)th entry of M^3. _____

_____

_____

_____

It can be verified that $M \times M^2 = M^3 = M^2 \times M$. But in general, matrix multiplication is not commutative. Matrix multiplication is associative, however, so

$$M \times M \times M = M \times M \times M$$
$$(M \times M) \times M = M \times (M \times M)$$
$$M^2 \times M = M \times M^2$$

b) $M^3(2,1) = 2$. This means that there are two three-edged paths from vertex 2 to vertex 1. One is

$$2\text{-}3,\ 3\text{-}4,\ 4\text{-}1.$$

What is the other?

$$2\text{-}\underline{},\ \underline{}\text{-}\underline{},\ \underline{}\text{-}1.$$

Now if we are interested in the total number of paths which link two vertices, we can compute the following matrix M^*,

$$M^* = M + M^2 + M^3 + \cdots + M^{K-1},$$

where K is the total number of vertices in the di-graph. That is, the (i,j)th element of M^* gives the total number of directed paths starting at vertex i and ending at vertex j. The total number of paths which join two vertices is of most interest in communication networks, where it is very desirable to have many ways that a message can travel from one vertex (transmitter) to another.

Let's compute M^* for the Communication Problem

$$M^* = M + M^2 + M^3.$$

$$\begin{bmatrix} 2 & 2 & 3 & 2 \\ 3 & 1 & 3 & 2 \\ 1 & 1 & 1 & 1 \\ 2 & 1 & 2 & 1 \end{bmatrix} = \begin{bmatrix} 0 & 1 & 1 & 0 \\ 1 & 0 & 1 & 0 \\ 0 & 0 & 0 & 1 \\ 1 & 0 & 0 & 0 \end{bmatrix} + \begin{bmatrix} 1 & 0 & 1 & 1 \\ 0 & 1 & 1 & 1 \\ 1 & 0 & 0 & 0 \\ 0 & 1 & 1 & 0 \end{bmatrix} + \begin{bmatrix} 1 & 1 & 1 & 1 \\ 2 & 0 & 1 & 1 \\ 0 & 1 & 1 & 0 \\ 1 & 0 & 1 & 1 \end{bmatrix}$$

From the above M^* matrix, we see that each vertex is reachable from every other vertex. However, many pairs of vertices (7 pairs) have only a single path which connects them. You will be asked to consider the implications of M^* in the Special Questions/Projects Section.

A museum is introducing "one-way" corridors for crowd control at popular exhibits. The proposed flow is shown below.

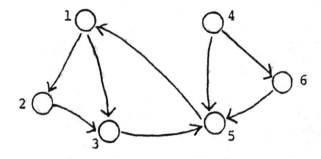

You, as the museum director, question whether it will be possible to visit all exhibits.

79. Derive M—the adjacency matrix of the di-graph.

To

	1 2 3 4 5 6
1	
2	
3	
4	
5	
6	

From

80. Calculate all matrices needed to check the existence of paths between vertices in this system. (The reader will need a separate sheet of paper.)

Please compare your results carefully with the Answers to Exercises since the next several questions are based on these matrices.

81. Calculate $M^* = M + M^2 + \cdots + M^{K-1}$.

82. How many paths exist:

   a) from vertex 6 to vertex 3? _____

   b) from vertex 2 to vertex 1? _____

83. Is every vertex reachable from every other vertex? _____

   Another interesting problem concerns the passing of toxins through the ecological food chain. For example, as DDT is applied, both grasslands and animals are exposed. The toxins are passed to other species according to the di-graph shown below.

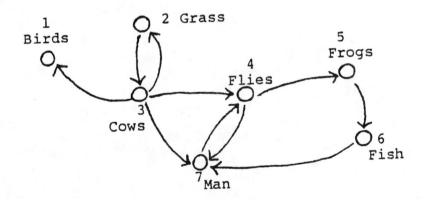

Ecologists follow the effects of DDT through no more than two intermediate species. Beyond two species, the effects are negligible.

84. What species are affected under such assumptions? (Are we again interested in computing $M^* = M + M^2 + \cdots + M^{K-1}$ or in some other matrix?)

Summary

If you have *worked* your way through this monograph, you are to be congratulated. A great deal of the information it contains is usually not presented until graduate school. You should be keenly aware of the many opportunities to apply graph theory to important problems. We hope this volume will serve as a useful reference as such problems come your way.

Appendix I
Definitions of Graph Theory Terms

Graph—A set of points or vertices, and a set of lines called edges which connect pairs of vertices.

Adjacent—Two vertices in a graph are said to be adjacent if they are joined by an edge.

Complete Graph—A graph is complete if every pair of distinct vertices is adjacent.

Bipartite Graph—A graph whose vertices can be separated into two distinct sets so that each edge has one endpoint in each set.

Degree—The degree of a vertex is the number of edges emanating from it.

Chromatic Number—The smallest number of "colors" required to color the vertices of a graph so that no two adjacent vertices have the same color.

Planarity—A graph is said to be planar if a picture of the graph can be drawn so that its edges intersect only at the vertices.

Connected—A graph is connected if a path of edges exists from any vertex to any other.

Cycle—A cycle is a path of edges, leading from a vertex to itself, which does not pass through any edge or vertex more than once.

Tree—A connected graph which contains no cycles.

Spanning Tree—A spanning tree of a graph G is a tree in G which contains every vertex of G.

Minimal Spanning Tree—A spanning tree of a weighted-edge graph G in which the sum of the weights is as small as possible.

Directed Graph—A graph in which there is a direction of flow indicated for each edge; also called a di-graph.

Critical Path—The longest path from the start-source to the end-sink in an activity network, that is, a di-graph showing durations and precedence relationships of activities within a project.

Reachability—The ability to get from one vertex to another in a directed graph.

Subgraph—If G^1 is a subgraph of G, then every vertex or edge of G^1 is a vertex or edge of G.

Source—In a di-graph, a vertex having outbound edges only.

Sink—In a di-graph, a vertex having inbound edges only.

Appendix II
A Biographical Sketch of
Leonhard Euler (1707–1783)

Euler was born in Basel, Switzerland, the son of a clergyman. His talent for mathematics was evident, and at an early age he was sent to the University of Basel. Studying under the famous John Bernoulli, he received the Master of Arts degree in mathematics, at the age of 17.

Euler traveled to St. Petersburg, in Russia, and took the Chair of Natural Philosophy in that University. It is said that some years later, while working on what was felt to be a critical but unsolvable problem, Euler lost the sight of one eye. Nonetheless, the problem was solved in three days!

Euler later moved to Germany. The esteem in which Euler was held is evident by the fact that once in 1760 a Russian army invaded Germany and pillaged a farm belonging to Euler. As soon as the attacking general realized whose farm had been ravaged he made immediate restitution.

In 1766 Euler returned to St. Petersburg where he shortly lost the sight of his other eye. Sutdents now copied down his words exactly as he dictated them. He developed an extraordinary capability to do extensive calculations in his head.

Euler made extensive contributions to graph theory, analysis, and the theory of numbers. In 1735 he solved the famous Königsberg Bridge Problem. He later proved that

$$V + F - E = 2$$

where V, F, and E are the numbers of vertices, faces, and edges of any simple polyhedron. "Let one small formula be quoted as the epitome of what Euler achieved:

$$e^{i\pi} + 1 = 0.$$

... Every symbol has its history—the principal whole numbers 0 and 1; the chief mathematical relations $+$ and $=$; π the discovery of Hippocrates; i the sign of the 'impossible' square root of minus one; and e the base of Napierian logarithms."*

---

*The World of Mathematics, edited by James R. Newman, Simon and Schuster, 1959, New York, pp. 150–151.

Appendix III
Matrices and Basic Matrix Operations

A *matrix* is a rectangular or square array of numbers. Matrices are used in mathematics, statistics, physics, economics, and other areas to represent many things. In this monograph matrices are used to represent graphs. Some examples of matrices are given below:

$$
\begin{array}{cccc}
A & B & C & D \\
\begin{bmatrix} 6 & 2 \\ 8 & 1 \end{bmatrix} &
\begin{bmatrix} 9 & 3 & 1 & 6 \\ 8 & 2 & 1 & 9 \\ 1 & 0 & 0 & 1 \end{bmatrix} &
\begin{bmatrix} 2 & 3 \\ 1 & 1 \end{bmatrix} &
\begin{bmatrix} 1 & 0 & 0 & 0 \\ 0 & 1 & 0 & 0 \\ 0 & 0 & 1 & 0 \\ 0 & 0 & 0 & 1 \end{bmatrix}
\end{array}
$$

Matrix A has 2 rows and 2 columns. Matrix B has 3 rows and 4 columns. A Matrix A is said to be "M by N", written $A_{M,N}$, if it has M rows and N columns.

1. What are the numbers of rows and columns for matrices C and D above?*

$$C\_\_,\_\_$$
$$D\_\_,\_\_$$

The elements of a matrix are usually numbers. The element in the ith row and the jth column of a matrix A is denoted $A(i,j)$. For example, in the matrices above, $A(1,1) = 6$ and $C(2,2) = 1$.

2. What is $B(2,2) =$
$D(3,2) =$
$A(2,1) =$

The addition and multiplication of matrices can have real physical meaning. Examples are given in the section on directed graphs. The intent of this appendix is to teach you the mechanics of those operations.

A. Matrix Addition

Two matrices A and B can be added to form a new matrix C only if A and B are the same size, that is, A and B may be added only if both are M by N. Matrix C which results is also M by N.

The elements of C are defined in the following way:

$$C(i,j) = A(i,j) + B(i,j),$$

---

*Answers to these problems appear in the Answer section.

that is, the element in the ith row and jth column of A is added to the element in the same location of B and that sum is the element in the ith row and jth column of C. Consider the following example:

$$\begin{array}{ccccc} A & + & B & = & C \\ \begin{bmatrix} 5 & 1 \\ 0 & 2 \end{bmatrix} & & \begin{bmatrix} 3 & 6 \\ 8 & 7 \end{bmatrix} & & \begin{bmatrix} & \\ & \end{bmatrix} \end{array}$$

Since A and B are both 2 by 2 they can be added to form matrix C, which will also have 2 rows and 2 columns.

The element in the first row and first column of C is by our definition:

$$C(1,1) = A(1,1) + B(1,1)$$
$$C(1,1) = 5 + 3 = 8$$

$$\begin{array}{ccccc} A & + & B & = & C \\ \begin{bmatrix} 5 & 1 \\ 0 & 2 \end{bmatrix} & & \begin{bmatrix} 3 & 6 \\ 8 & 7 \end{bmatrix} & & \begin{bmatrix} 8 & \underline{} \\ \underline{} & \underline{} \end{bmatrix} \end{array}$$

3. Compute the remaining elements of C.

$$\begin{array}{c} C \\ 8 \quad \underline{} \\ \underline{} \quad \underline{} \end{array}$$

4. Compute the sum of the following two matrices.

$$\begin{array}{ccccc} A & + & B & = & C \\ \begin{bmatrix} 3 & 1 & 0 \\ 0 & 2 & 6 \end{bmatrix} & & \begin{bmatrix} 2 & 2 & 1 \\ 5 & 8 & 6 \end{bmatrix} & & \begin{bmatrix} \underline{} & \underline{} & \underline{} \\ \underline{} & \underline{} & \underline{} \end{bmatrix} \end{array}$$

5. Compute the sum of matrices B and D from page 55.

$$\begin{array}{ccccc} B & + & D & = & C \\ \begin{bmatrix} 9 & 3 & 1 & 6 \\ 8 & 2 & 1 & 9 \\ 1 & 0 & 0 & 1 \end{bmatrix} & & \begin{bmatrix} 0 & 1 & 0 & 0 \\ 0 & 1 & 0 & 0 \\ 0 & 0 & 1 & 0 \\ 0 & 0 & 0 & 1 \end{bmatrix} & & \end{array}$$

B. Matrix Multiplication

Matrix $A_{M,N}$ can be multiplied by any matrix B which has as many rows as A has columns, that is, the product

$$A_{M,N} \times B_{N,P}$$

can be computed. But the product

$$A_{M,N} \times B_{R,P}$$

cannot be computed unless $R = N$. For example, $A_{2,3}$ can be multiplied by $B_{3,4}$, but not by $B_{4,4}$. $A_{3,6}$ can be multiplied by $B_{6,K}$ where K is any positive integer.

6. Consider $A_{2,2}$ and $B_{2,3}$; can they be multiplied? _____

7. What will be the size of $A_{2,2} \times B_{2,3} = C_{\_,\_}$?

When two matrices $A_{M,N}$ and $B_{N,K}$ are multiplied, the resulting matrix C is M by K. But the question is how do we compute $C(i,j)$, the elements of C.

$C(i,j)$ is found by individually multiplying the elements in the ith row of A by those in the jth column of B and adding those products.

An example will make the procedure much clearer than words.

The method of computing an element of C will now be shown. From our definition, $C(1,1)$ is found by individually multiplying the elements of the first row of A by those in the first column of B and adding the products.

$$
\begin{array}{ccccc}
A & \times & B & = & C \\
\begin{bmatrix} 5 & 4 \\ 3 & 2 \end{bmatrix} & & \begin{bmatrix} 2 & 1 & 2 \\ 0 & 4 & 3 \end{bmatrix} & & \begin{bmatrix} 1,1 & \underline{} & \underline{} \\ \underline{} & \underline{} & \underline{} \end{bmatrix}
\end{array}
$$

That is,
$$ C(1,1) = 5 \cdot 2 + 4 \cdot 0 = 10. $$

$C(2,2)$ is found by individually multiplying the elements of the second row of A by the elements in the second column of B and adding the products.

$$
\begin{array}{ccccc}
A & \times & B & = & C \\
\begin{bmatrix} 5 & 4 \\ 3 & 2 \end{bmatrix} & & \begin{bmatrix} 2 & 1 & 2 \\ 0 & 4 & 3 \end{bmatrix} & & \begin{bmatrix} 10 & \underline{} & \underline{} \\ \underline{} & 2,2 & \underline{} \end{bmatrix}
\end{array}
$$

That is,
$$ C(2,2) = 3 \cdot 1 + 2 \cdot 4 = 11. $$

$$
\begin{array}{ccccc}
A & \times & B & = & C \\
\begin{bmatrix} 5 & 4 \\ 3 & 2 \end{bmatrix} & & \begin{bmatrix} 2 & 1 & 2 \\ 0 & 4 & 3 \end{bmatrix} & & \begin{bmatrix} 10 & \underline{} & \underline{} \\ \underline{} & 11 & \underline{} \end{bmatrix}
\end{array}
$$

8. Compute the remaining elements of C and write them in the spaces above.

9. For the following pairs of matrices, compute, if possible, their product:

a)
$$
\begin{array}{ccccc}
A & \times & B & = & C \\
\begin{bmatrix} 6 & 2 \\ 8 & 1 \end{bmatrix} & & \begin{bmatrix} 2 & 3 \\ 1 & 1 \end{bmatrix} & & \begin{bmatrix} \end{bmatrix}
\end{array}
$$

b)
$$
\begin{array}{ccccc}
A & \times & B & = & C \\
\begin{bmatrix} 6 & 5 \\ 8 & 8 \end{bmatrix} & & \begin{bmatrix} 2 & 6 & 9 \\ 8 & 2 & 1 \\ 1 & 1 & 1 \end{bmatrix} & & \begin{bmatrix} \end{bmatrix}
\end{array}
$$

c)
$$
\begin{array}{ccccc}
A & \times & B & = & C \\
\begin{bmatrix} 9 & 3 & 1 & 6 \\ 8 & 2 & 1 & 9 \\ 1 & 0 & 0 & 1 \end{bmatrix} & & \begin{bmatrix} 1 & 0 & 0 & 0 \\ 0 & 1 & 0 & 0 \\ 0 & 0 & 1 & 0 \\ 0 & 0 & 0 & 1 \end{bmatrix} & & \begin{bmatrix} \end{bmatrix}
\end{array}
$$

Compare matrix C and matrix A in 9c). What do you observe? (Matrix B is called an identity matrix.) Can you guess why?

A final note. While multiplication of real numbers is commutative, that is $6 \times 3 = 3 \times 6$, multiplication of matrices is not commutative.

10. **a)** Compute

$$
\begin{array}{ccccc}
B & \times & A & = & C \\
\begin{bmatrix} 2 & 3 \\ 1 & 1 \end{bmatrix} & & \begin{bmatrix} 6 & 2 \\ 8 & 1 \end{bmatrix} & & \begin{bmatrix} & \\ & \end{bmatrix}
\end{array}
$$

Does this equal the answer to 9a)?

10. **b)**

$$
\begin{array}{ccccc}
A & \times & B & = & C \\
\begin{bmatrix} 1 & 0 & 0 & 0 \\ 0 & 1 & 0 & 0 \\ 0 & 0 & 1 & 0 \\ 0 & 0 & 0 & 1 \end{bmatrix} & & \begin{bmatrix} 9 & 3 & 1 & 6 \\ 8 & 2 & 1 & 9 \\ 1 & 0 & 0 & 1 \end{bmatrix} & & \begin{bmatrix} & & & \\ & & & \end{bmatrix}
\end{array}
$$

Does this equal the answer to 9c)?

Special Questions/Projects

1. Derive a computer routine for finding the chromatic number of a graph.

2. Prove that a tree with N vertices must have exactly $N - 1$ edges.

3. Develop computer routines for:

Kruskal's Algorithm

Dijkstra's Algorithm

Matrix Multiplication

Answering Reachability Problems

4. In the section on reachability, communication networks were discussed. In particular, question 74 asks you to define the matrix representation of such a network. If one were attempting to sabotage this network, what single link, if removed, would have the most catastrophic effect on the network?

5. A trucking firm located in New York City delivers goods to a number of cities in the United States and Canada. The cities and driving times between cities appear in the following figure. The firm is interested in determining the shortest driving time from New York to each city the trucks visit. Use the appropriate algorithm to obtain the times.

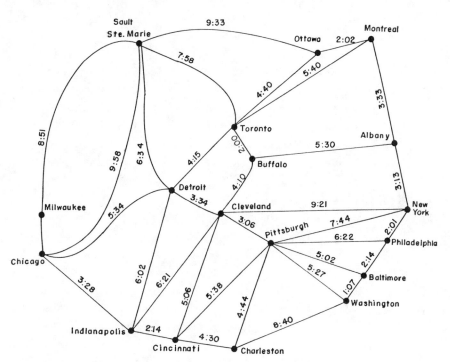

6. Suppose you are a PCB "builder." An electronics engineer provides you with a complicated circuit which is to be built on a "board" and claims that Kuratowski's theorem indicates the graph is planar. How would you attempt to find a planar representation of the circuit which could be used as a model for the PCB?

7. Write a computer program which takes as input a matrix description of a graph and uses the result of Kuratowski's theorem to determine if the graph is planar.

References

1. *Discrete Mathematical Structures for Computer Science*, R. E. Prather, Houghton Mifflin Co., Boston, MA, 1976 (chapter 2).

2. *Graphs and Their Uses*, O. Ore, Random House, New York, NY, 1963.

3. *Graphs, Models and Finite Mathematics*, J. Malkevitch and W. Meyer, Prentice Hall, Inc., Englewood Cliffs, NJ, 1974.

4. *Introduction to Operations Research*, F. S. Hillier and G. J. Lieberman, Holden-Day, Inc., San Francisco, CA, 1967 (chapter 7).

5. *The World of Mathematics*, J. R. Newman, Simon and Schuster, New York, NY, 1956.

6. *Application of a Technique for Research and Development Program Evaluation*, D. G. Malcolm, et al., Operations Research 7: 646–669, 1959.

7. *Network Analysis for Management Decisions*, S. M. Lee, et al., Kluwer-Nijhoff Publishing Co., Boston, MA, 1982.

8. *Dots and Lines*, R. J. Trudeau, Kent State University Press, Kent State University, 1976.

9. *Graph Theory with Applications to Engineering and Computer Science*, Narsingh Deo, Prentice-Hall, Englewood Cliffs, NJ, 1974.

Answers

Answers to Exercises

1.

Number of:	Vertices	Edges
Graph C	6	8
Graph D	4	6
Graph F	5	6
Graph H	10	17

2. (a) 1–2, 1–5, 2–3, 3–4, 3–5, 3–6, 4–6, 5–6

(b) 1–3, 1–4, 1–6, 2–4, 2–5, 2–6, 4–5

3. D, E, G

4. (a) 6

(b) 10

(c) $N(N-1)/2$—Which is seen by the following argument: In a complete graph each of the N vertices has $N-1$ edges which connect that vertex to each of the remaining ones. But this method counts each edge twice, since every edge is associated with 2 vertices. Hence, the divisor of 2 is required.

5. (a) Yes $\{1, 5, 4\}$; $\{2, 6, 3\}$

(b) Yes $\{9\}$; $\{1, 2, 3, 4, 5, 6, 7, 8\}$

(c) No—try to construct the two necessary disjoint sets.*

---

*Two sets with no elements in common are said to be *disjoint*.

(d) Yes $\{1, 3, 5\}$; $\{2, 4, 6\}$

6. (a) 9

(b) 15

(c) MN

7. $K_{5,3}$

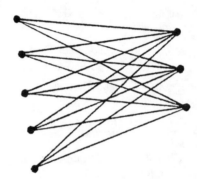

8.

Vertex	1	2	3	4	5	6	7	8	9	10
Degrees	3	3	3	4	2	4	3	2	7	3

9. (a) 3, 4, 6 **(b)** 4

10. Chromatic Number

(a) 2

(b) 2

(c) 4

(d) 2

(e) 2

(f) 3

11. (a) 1

(b) N—Since in a complete graph every vertex is adjacent to every other vertex. (Does it also follow that every graph with N

vertices and chromatic number *N* must be complete?)

(c) 2

(d) 2—Use one color for all vertices in each of the two disjoint sets.

12. If *any* student were taking *both* subjects, their finals would have to be scheduled at different times.

13. If *no* student is taking both courses, the two finals could be scheduled at the same time.

14. No

15. Only that some student is taking both courses.

16.

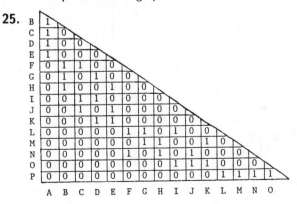

17. 4; A coloring is:
∘ = {3, 5, 7}
△ = {2, 4, 8}
□ = {6, 10}
◇ = {1, 9}

Other "four colorings" are possible.

18. No—by the definition of "coloring."

19. Subjects colored the same can hold their final exams at the same time.

20. No

21. There is at least 1 student, #8, who is taking 4 subjects.

22. 3

23.

Area	"Color"	Chemicals
1	∘	1, 2, 4, 7
2	△	5, 6
3	□	8, 3, 9, 10

Your answer may be correct even though it does not match the cue given above. There are several possible "colorings" which use three colors, another is:

Area	"Color"	Chemicals
1	□	1, 3, 5, 10
2	△	4, 6, 7, 8, 9
3	∘	2

24. Note that chemicals 2, 3, and 6 are all incompatible with one another, hence at *least* 3 colors are required for the graph.

25.

	A	B	C	D	E	F	G	H	I	J	K	L	M	N	O
B	1														
C	1	0													
D	1	0	0												
E	1	0	0	0											
F	0	1	1	0	0										
G	0	1	0	1	0	0									
H	0	1	0	0	1	0	0								
I	0	0	1	1	0	0	0	0							
J	0	0	1	0	1	0	0	0	0						
K	0	0	0	1	1	0	0	0	0	0					
L	0	0	0	0	0	1	1	0	1	0	0				
M	0	0	0	0	0	0	1	1	0	0	1	0			
N	0	0	0	0	0	1	0	1	0	1	0	0	0		
O	0	0	0	0	0	0	0	0	1	1	1	0	0	0	
P	0	0	0	0	0	0	0	0	0	0	0	1	1	1	1

26. (a)

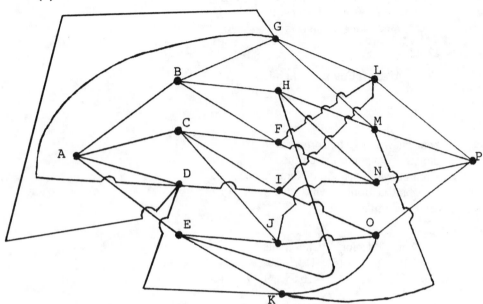

(b) There are 2 possible sets of 8 letters,
{A, G, H, F, I, J, K, P} and
{B, C, D, E, L, M, N, O}.

27. 2

28. No. The following two sets each contain 8 nonconfusable letters.

□ = {B, C, D, E, L, M, N, O}
∘ = {A, F, G, H, I, J, K, P}

29.

	1	2	3	4
1		1	0	1
2	1		0	0
3	0	0		1
4	1	0	1	

30. If there is (or is not) an edge between vertices i and j there is (or is not) an edge between vertices j and i.

31. (a)

2	0	
3	0	1
	1	2

(b)

2	1				
3	0	1			
4	0	0	1		
5	1	0	1	0	
6	0	0	1	1	1
	1	2	3	4	5

(c)

2	1		
3	1	1	
4	1	1	1
	1	2	3

(d)

2	1		
3	1	1	
4	1	1	1
	1	2	3

Note that this is K_2—a complete graph.

(e)

2	0			
3	0	0		
4	1	1	1	
5	1	1	1	0
	1	2	3	4

32. A coloring and the largest independent set are:

$\square$ = {PUL, SBP, LIPC, BVR, GCS}—largest independent set;

○ = {PSTR, PUPL};

△ = {CAPR, BMR};

◇ = {PUPR};

▲ = {PSL};

◆ = {PSR}

Did you note that BVR, BMR, PUPL, PUPR, PSL and PSR form K_6?

Thus the chromatic number of the graph must be at least 6.

33. No—it's impossible.

34. A complete Bipartite Graph with sets of 3 and 3 vertices, $K_{3,3}$.

35. **(a)** yes

(b) yes—Of course it's true that *as drawn* the graph does *not* appear to be planar. But recall that there are many pictures of any graph and that to show a graph is planar only one "planar" picture of it needs to be drawn. Consider the following picture:

This graph has the same four vertices as 35(b). You can also verify that it has the same edges as 35(b). Therefore, it's just another picture of graph 35(b). Since this picture is planar, the graph of 35(b) is planar. It is only the picture of that graph given in the text which is not planar.

(c) yes

(d) yes—There does exist a planar picture of this graph.

(e) no

36. **(a)** K_4 Planar

(b) K_6 Nonplanar

(c) $K_N, N \geq 5$ Nonplanar

(d) Cube   Planar

Kuratowski's Theorem—K_6 Contains K_5 as a subgraph.

Kuratowski's Theorem—$K_N, N \geq 5$ Contains K_5 as a subgraph.

37. A, C, D, E, F, G, H

38. C

A is not a tree because it contains a cycle:

B is not a tree because it is not connected.

39.

 ; ; ; ;

Note that other possibilities do exist.

40. A and D: A to C to D

B and F: B to A to C to D to F

C and G: C to D to E to G

41.

42. Edge

1. B to F
2. C to E
3. A to C
4. B to E
5. C to D

The resulting minimal spanning tree is:

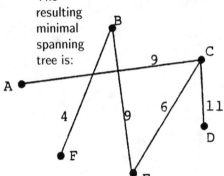

43.

	x_1	x_2	x_3	x_4
Patient 7	5	8	14	10
Patient 15	10	10	3	12
Patient 23	3	2	1	12

44.

	D
Patients 7 and 12	24
Patients 9 and 34	5
Patients 10 and 25	3

45.

	D
Patients 4 and 23	31
Patients 13 and 35	30
Patients 10 and 20	25

46. These 3 pairs of points are at distance zero from one another.

47. No

48. No

49. Well, does it?

50.

Travel from U_0 to	Vertices Encountered	Minimal Cost of Path
A	A	6
B	B	3
C	C	2
D	A, D	8
E	A, D, E	9

51. No

52. B to C, B to D, and C to E

53.

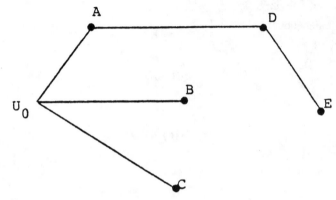

54. It is a tree.

55.

Vertex	Path from U_0	Dist.
A	U_0A	3
B	U_0AB	7
C	U_0ABC	8
D	U_0D	8
E	U_0ABE	13
F	U_0ABEF	17
G	U_0DG	13
H	U_0ABEH	15
I	U_0ABEHI	16

56. No flow leaves vertex 2.

57 (a) A "1" indicating an edge *from* the sink *to* itself.

(b) 1

(c) 1—The in-degree of a vertex may be found by

counting the "1"'s in the "To" column of the vertex in the matrix representation. For the sample graph:

	To		
	1	2	3
From 1	1	1	1
2	0	0	0
3	1	1	0

there is a single entry in the "To 3" Column—namely for the edge coming from vertex 1. Thus the in-degree of vertex 3 is 1.

58. (a)

	To				
	1	2	3	4	5
From 1	0	1	1	0	0
2	0	0	0	0	0
3	0	0	1	0	0
4	0	0	1	0	1
5	0	0	0	1	1

(b)

	To					
	1	2	3	4	5	6
From 1	0	1	0	1	0	0
2	0	0	1	0	0	0
3	0	0	0	0	1	0
4	0	1	0	0	1	0
5	0	0	1	0	0	1
6	0	0	0	0	0	0

59. (a)

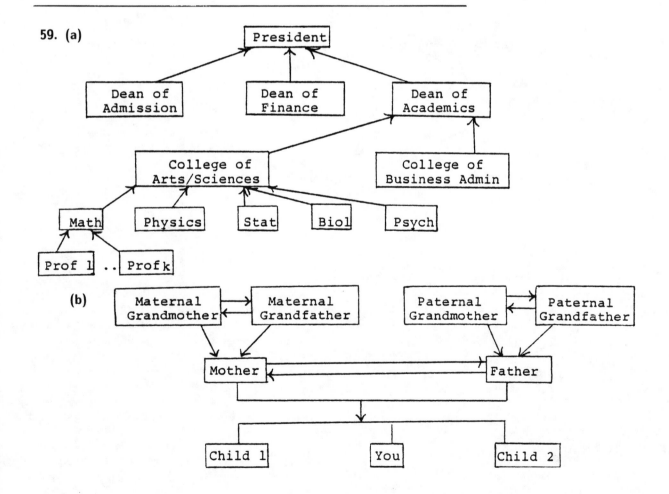

(b)

60. 26 Minutes

61.

Path No.	Activity Sequence	Time from S to F
1	S – T – E – F	$0 + 3 + 9 + 0 = 12$
2	S – C – E – F	$0 + 4 + 9 + 0 = 13$
3	S – ST – E – F	$0 + 2 + 9 + 0 = 11$
4	S – FS – FE – E – F	$0 + 13 + 4 + 9 + 0 = 26$

62.

Event	Earliest Time	Latest Time	Slack Time
FS	0	0	0
ST	0	15	15
C	0	13	13
T	0	14	14
FE	13	13	0
E	17	17	0

63. Events which lie on the critical path have *no* slack time.

64.

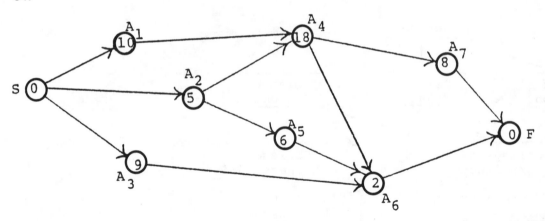

65. 6

66.

Path No.	Activity Sequence	Time from S to F
1	S – A_1 – A_4 – A_7 – F	36
2	S – A_1 – A_4 – A_6 – F	30
3	S – A_2 – A_4 – A_7 – F	31
4	S – A_2 – A_4 – A_6 – F	25
5	S – A_2 – A_5 – A_6 – F	13
6	S – A_3 – A_6 – F	11

67. 36

68. A_1, A_4, A_7

69.

Event	Earliest Time	Latest Time	Slack Time
A_1	0	0	0
A_2	0	5	5
A_3	0	25	25
A_4	10	10	0
A_5	5	28	23
A_6	28	34	6
A_7	28	28	0

70. $A_1, A_4, A_7, A_2, A_5, A_6, A_3$

71.

Activity	Immediately Preceding Activities
A_1	S
A_2	A_1
A_3	A_2
A_4	S, A_1
A_5	A_4
A_6	A_2, A_5
A_7	S
A_8	A_7

72.

Path No.	Activity Sequence	Time from S to F
1	$S - A_1 - A_2 - A_3 - F$	13
2	$S - A_1 - A_2 - A_6 - F$	16
*3	$S - A_1 - A_4 - A_5 - A_6 - F$	25
4	$S - A_7 - A_8 - F$	11
5	$S - A_4 - A_5 - A_6 - F$	22

73.

Event	Earliest Time	Latest Time	Slack Time	Management Priority
A_1	0	0	0	1
A_2	3	12	9	2
A_3	4	16	12	3
A_4	3	3	0	1
A_5	5	5	0	1
A_6	13	13	0	1
A_7	0	14	14	4
A_8	5	19	14	4

74.

	To			
From	1	2	3	4
1	0	1	1	0
2	1	0	1	0
3	0	0	0	1
4	1	0	0	0

75.

	To			
M^2	1	2	3	4
1	1	0	1	1
2	0	1	1	1
3	1	0	0	0
4	0	1	1	0

From (rows 1–4)

76. (a) 1

(b) 1

77.

	To			
M^3	1	2	3	4
1	1	1	1	1
2	2	0	1	1
3	0	1	1	0
4	1	0	1	1

From (rows 1–4)

78. (a) Element (3, 4) of M^3 is zero. Thus, there are no 3–edged paths from vertex 3 to vertex 4.

(b) 2-1, 1-2, 2-1.

79.

	To					
M	1	2	3	4	5	6
1	0	1	1	0	0	0
2	0	0	1	0	0	0
3	0	0	0	0	1	0
4	0	0	0	0	1	1
5	1	0	0	0	0	0
6	0	0	0	0	1	0

From (rows 1–6)

80.

	To					
M^2	1	2	3	4	5	6
1	0	0	1	0	1	0
2	0	0	0	0	1	0
3	1	0	0	0	0	0
4	1	0	0	0	1	0
5	0	1	1	0	0	0
6	1	0	0	0	0	0

From (rows 1–6)

	To					
M^3	1	2	3	4	5	6
1	1	0	0	0	1	0
2	1	0	0	0	0	0
3	0	1	1	0	0	0
4	1	1	1	0	0	0
5	0	0	1	0	1	0
6	0	1	1	0	0	0

From (rows 1–6)

	To					
M^4	1	2	3	4	5	6
1	1	1	1	0	0	0
2	0	1	1	0	0	0
3	0	0	1	0	1	0
4	0	1	2	0	1	0
5	1	0	0	0	1	0
6	0	0	1	0	1	0

From (rows 1–6)

	To					
M^5	1	2	3	4	5	6
1	0	1	2	0	1	0
2	0	0	1	0	1	0
3	1	0	0	0	1	0
4	1	0	1	0	2	0
5	1	1	1	0	0	0
6	1	0	0	0	1	0

From (rows 1–6)

81.

	To					
M^*	1	2	3	4	5	6
1	2	3	5	0	3	0
2	1	1	3	0	2	0
3	2	1	2	0	3	0
4	3	2	4	0	5	1
5	3	2	3	0	2	0
6	2	1	2	0	3	0

From (rows 1–6)

82. (a) 2

(b) 1

83. No. There are no paths by which vertex 4 can be reached from any other vertex. Vertex 6 can only be reached by vertex 4-note the single "1" in the "To" column for that vertex.

84. Grass and man are those species initially exposed. The question asks you to identify the species which are affected by starting from grass or man and passing through no more than two intermediate species. Thus, we must determine M and compute M^2 and M^3 and add the three. Our answer is found by counting the nonzero column entries of the resulting matrix in the rows corresponding to man and grass.

Let – Birds = 1 Frogs = 5
Grass = 2 Fish = 6
Cows = 3 Man = 7
Flies = 4

	To						
M	1	2	3	4	5	6	7
1	0	0	0	0	0	0	0
2	0	0	1	0	0	0	0
3	1	1	0	1	0	0	1
From 4	0	0	0	0	1	0	1
5	0	0	0	0	0	1	0
6	0	0	0	0	0	0	1
7	0	0	0	1	0	0	0

	To						
M²	1	2	3	4	5	6	7
1	0	0	0	0	0	0	0
2	1	1	0	1	0	0	1
3	0	0	1	1	1	0	1
From 4	0	0	0	1	0	1	0
5	0	0	0	0	0	0	1
6	0	0	0	1	0	0	0
7	0	0	0	0	1	0	1

	To						
M³	1	2	3	4	5	6	7
1	0	0	0	0	0	0	0
2	0	0	1	1	1	0	1
3	1	1	0	2	1	1	2
From 4	0	0	0	0	1	0	2
5	0	0	0	1	0	0	0
6	0	0	0	0	1	0	1
7	0	0	0	1	0	1	0

By adding the second and seventh rows from these matrices (corresponding to grass and man) we obtain:

	1	2	3	4	5	6	7
(2) Grass	1	1	2	2	1	0	2
(7) Man	0	0	0	2	1	1	1

Thus, we see that every species considered is *reachable* in no more than 2 steps by starting with affected grass or man.

Answers to Questions in Appendix III

1. C2, 2; D4, 4

2. B(2, 2) = 2; D(3, 2) = 0; A(2, 1) = 8:

3.
$$\begin{array}{c} C \\ \begin{bmatrix} 8 & 7 \\ 8 & 9 \end{bmatrix} \end{array}$$

4.
$$\begin{array}{c} C \\ \begin{bmatrix} 5 & 3 & 1 \\ 5 & 10 & 12 \end{bmatrix} \end{array}$$

5. The sum *cannot* be computed. The matrices are of different size.

6. Yes

7. C(2, 3)

8.
$$\begin{array}{c} C \\ \begin{bmatrix} 10 & 21 & 22 \\ 6 & 11 & 12 \end{bmatrix} \end{array}$$

9. (a)
$$\begin{matrix} & C & \\ \begin{bmatrix} 14 & 20 \\ 17 & 25 \end{bmatrix} \end{matrix}$$

(b) C

Cannot be computed—matrices not of proper size.

(c)
$$\begin{matrix} & & C & & \\ \begin{bmatrix} 9 & 3 & 1 & 6 \\ 8 & 2 & 1 & 9 \\ 1 & 0 & 0 & 1 \end{bmatrix} \end{matrix}$$

10. (a)
$$\begin{matrix} & C & \\ \begin{bmatrix} 36 & 7 \\ 14 & 3 \end{bmatrix} \end{matrix}$$

No.

(b) Cannot be computed